TRUE SURVIVAL GRAPHICS

STRANDED at SEA

by Jarrod Luján   illustrated by Eduardo Gutiérrez

CAPSTONE PRESS
a capstone imprint

Published by Capstone Press, an imprint of Capstone
1710 Roe Crest Drive, North Mankato, Minnesota 56003
capstonepub.com

Copyright © 2024 by Capstone. All rights reserved. No part of this publication may be reproduced in whole or in part, or stored in a retrieval system, or transmitted in any form or by any means, electronic, mechanical, photocopying, recording, or otherwise, without written permission of the publisher.

Library of Congress Cataloging-in-Publication Data is available on the Library of Congress website.

ISBN: 9781669058717 (hardcover)
ISBN: 9781669058793 (paperback)
ISBN: 9781669058809 (ebook PDF)

Summary: Three teens run out of gas while island-hopping in the South Pacific. A lone sailor drifts on an inflatable life raft in the vast Atlantic. Four men cling to their capsized sailboat far off the New Zealand coast. These remarkable true tales could have ended in tragedy—but they didn't! What happened when these people found themselves stranded at sea for weeks on end? And how did they survive to tell their tales?

Editorial Credits
Editor: Christopher Harbo; Designer: Tracy Davies;
Production Specialist: Katy LaVigne

All internet sites appearing in back matter were available and accurate when this book was sent to press.

TABLE OF CONTENTS

Introduction
EARTH'S MIGHTY SEAS! 4

Chapter 1
TROUBLE IN TOKELAU! 6

Chapter 2
SO LONG, SOLO! 16

Chapter 3
DISASTER DOWN UNDER! 28

MORE ABOUT THESE TALES OF SURVIVAL 44

GLOSSARY 46

INTERNET SITES 47

BOOKS IN THE SERIES 47

ABOUT THE AUTHOR 48

ABOUT THE ILLUSTRATOR 48

INTRODUCTION

EARTH'S MIGHTY SEAS!

As humans, we depend on the Earth's seas.

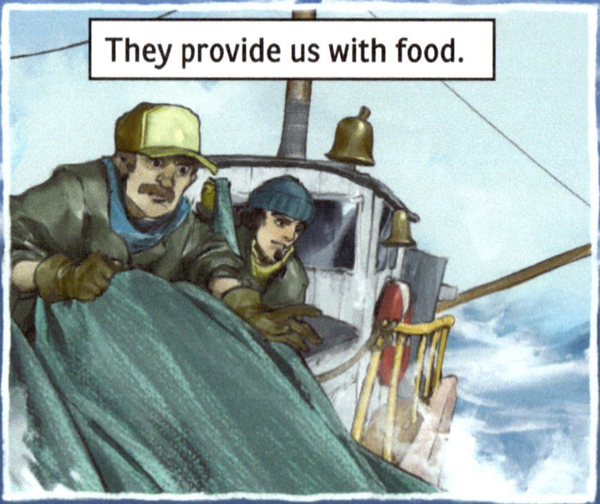
They provide us with food.

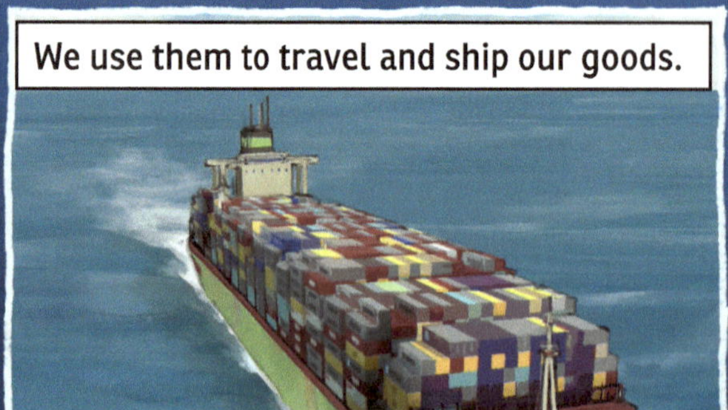
We use them to travel and ship our goods.

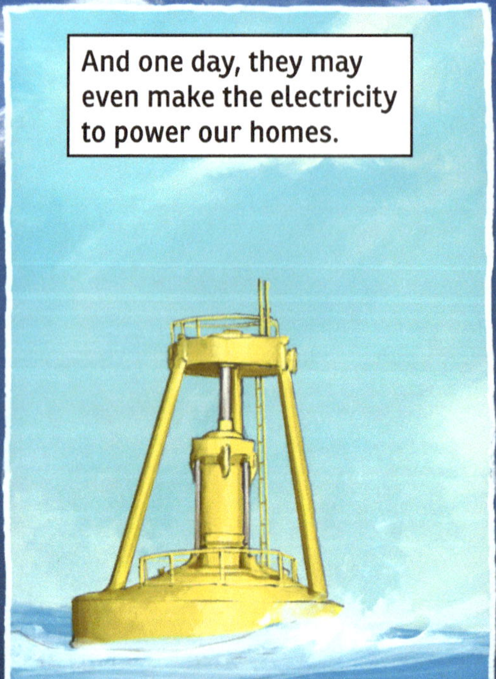
And one day, they may even make the electricity to power our homes.

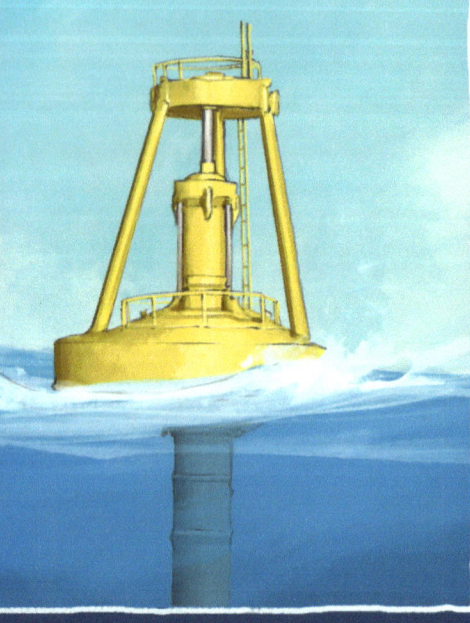

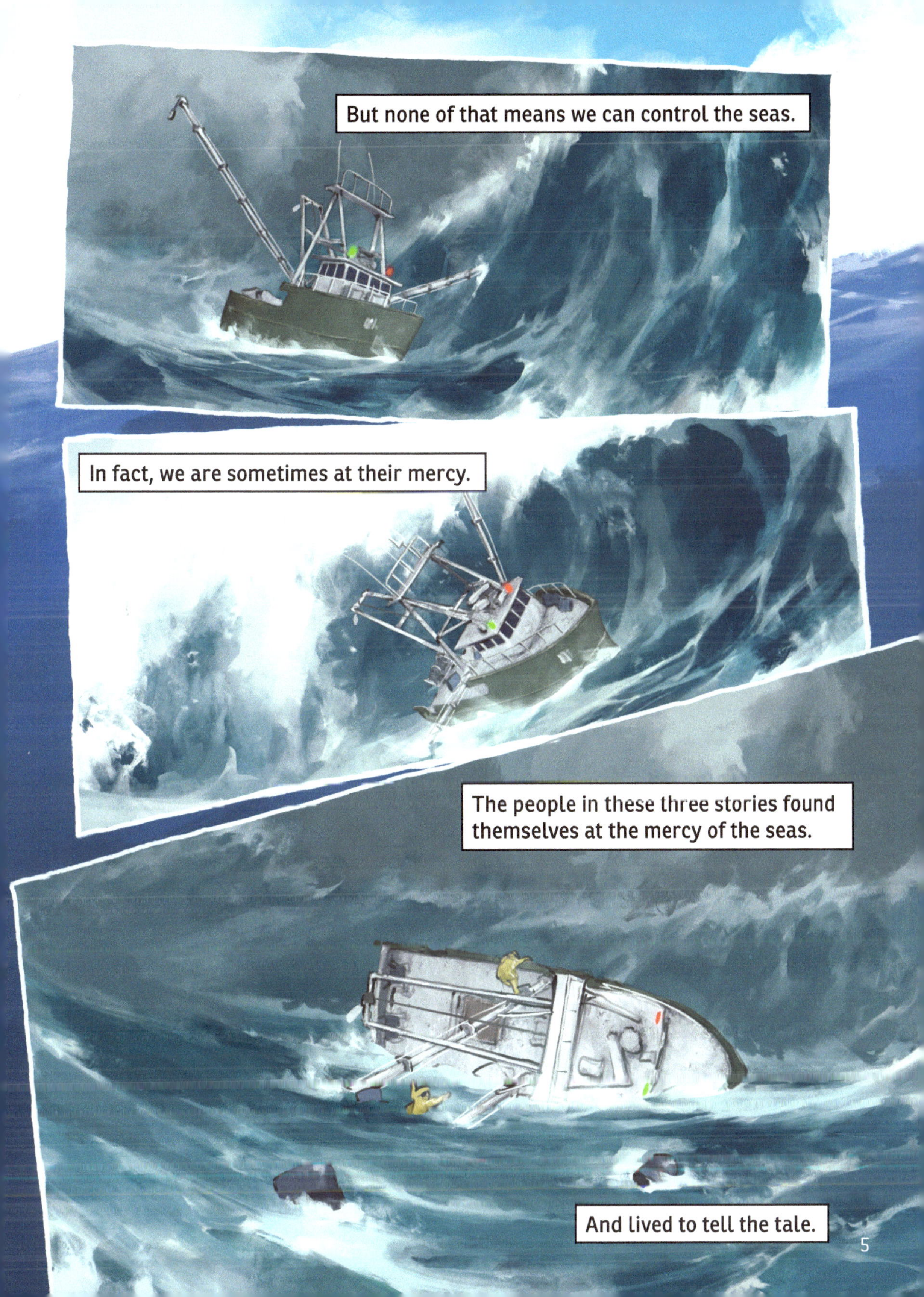
But none of that means we can control the seas.
In fact, we are sometimes at their mercy.
The people in these three stories found themselves at the mercy of the seas.
And lived to tell the tale.

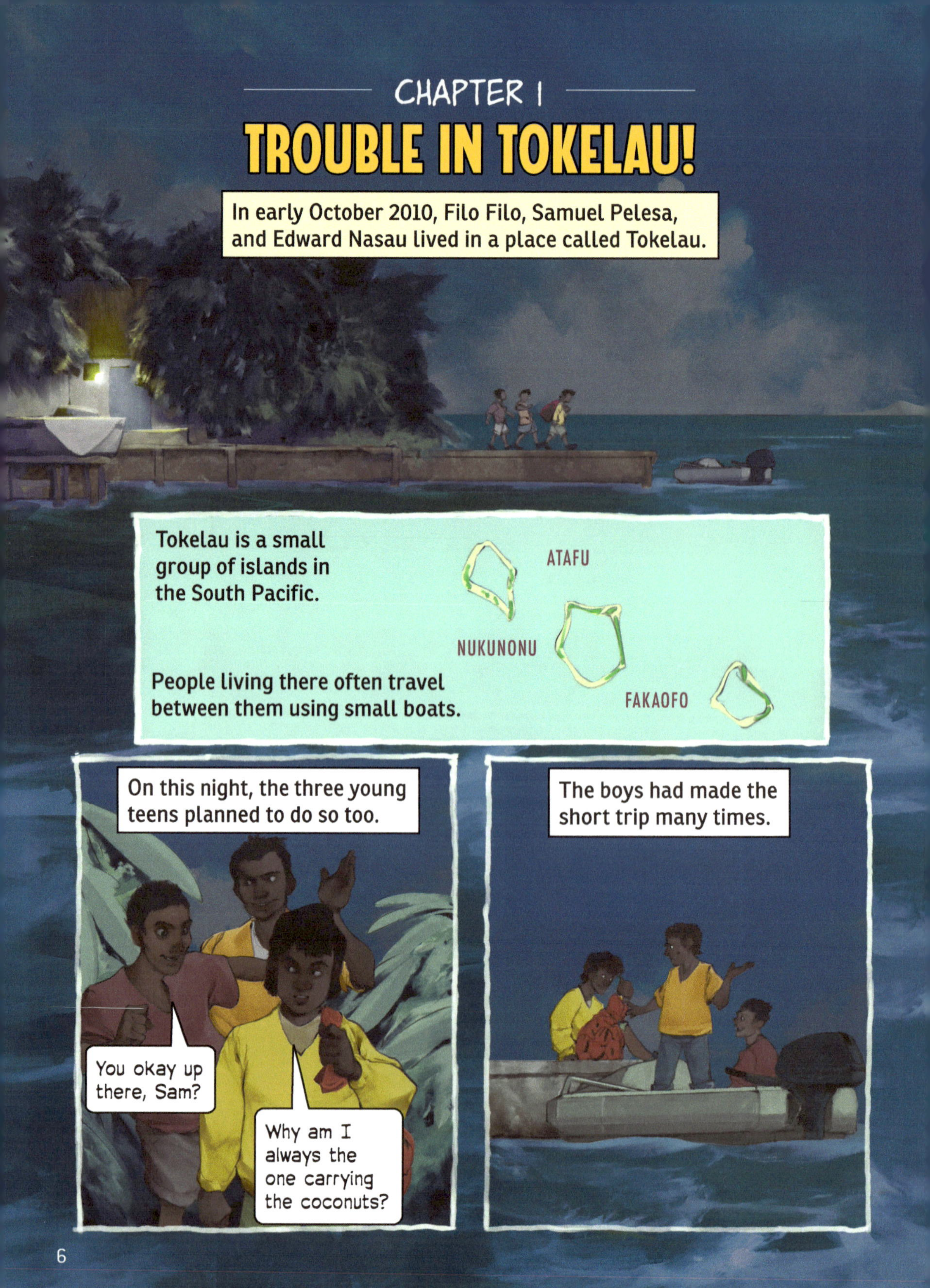
CHAPTER 1
TROUBLE IN TOKELAU!
In early October 2010, Filo Filo, Samuel Pelesa, and Edward Nasau lived in a place called Tokelau.
Tokelau is a small group of islands in the South Pacific.
ATAFU
NUKUNONU
FAKAOFO
People living there often travel between them using small boats.
On this night, the three young teens planned to do so too.
You okay up there, Sam?
Why am I always the one carrying the coconuts?
The boys had made the short trip many times.

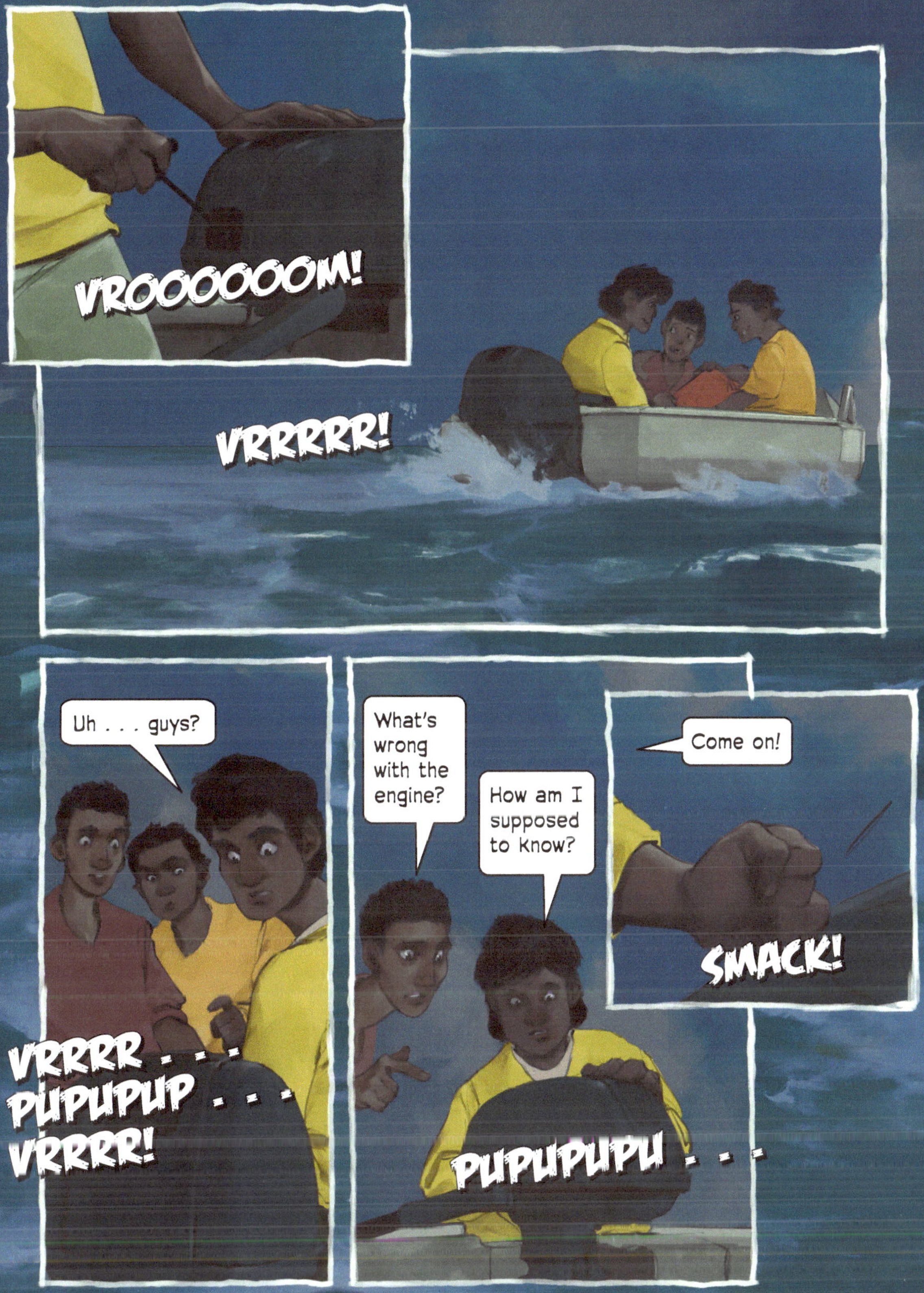
VROOOOOOM!
VRRRRR!
Uh . . . guys?
VRRRR . . .
PUPUPUP . . .
VRRRR!
What's wrong with the engine?
How am I supposed to know?
PUPUPUPU . . .
Come on!
SMACK!

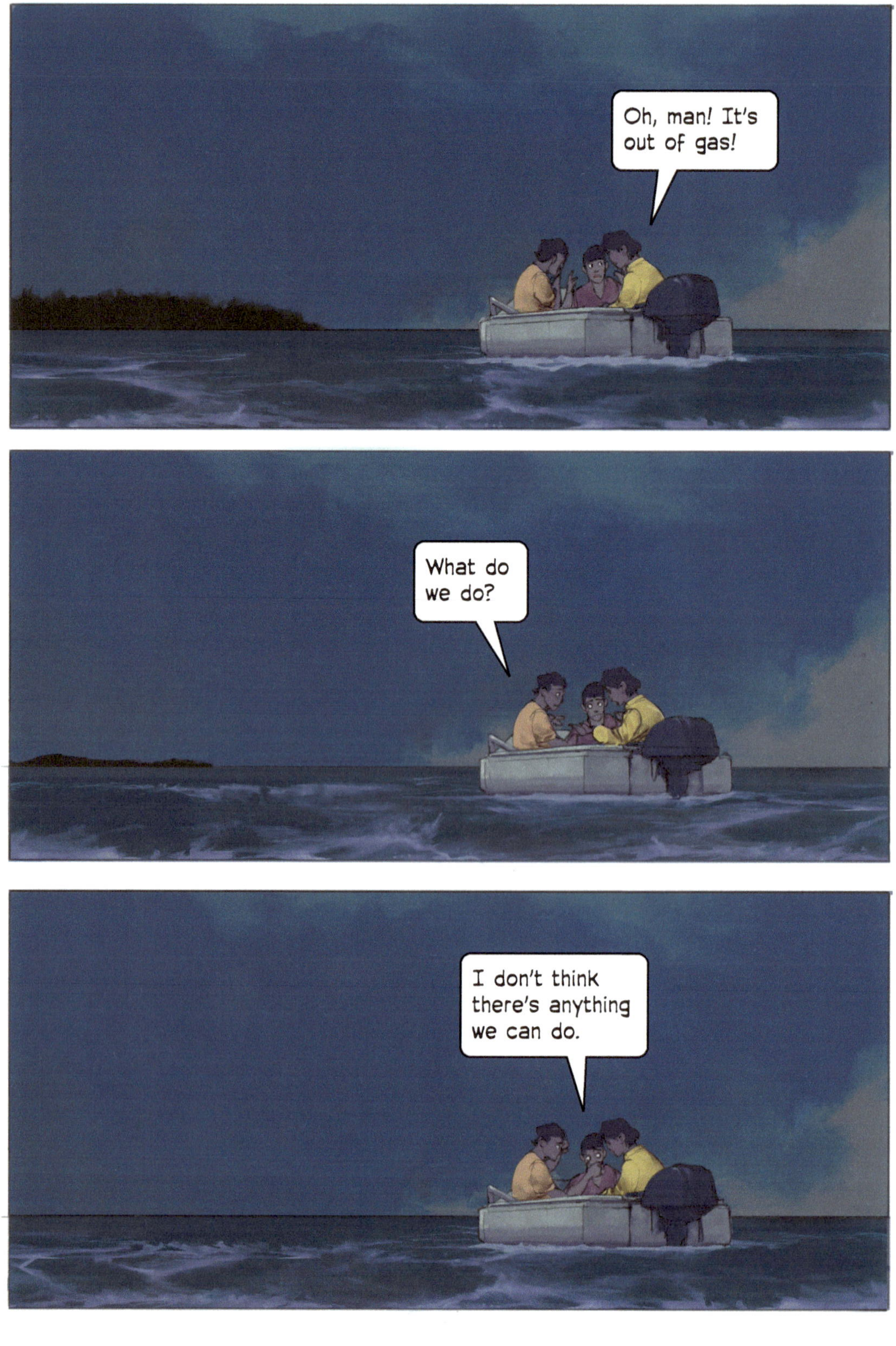
Oh, man! It's out of gas!
What do we do?
I don't think there's anything we can do.

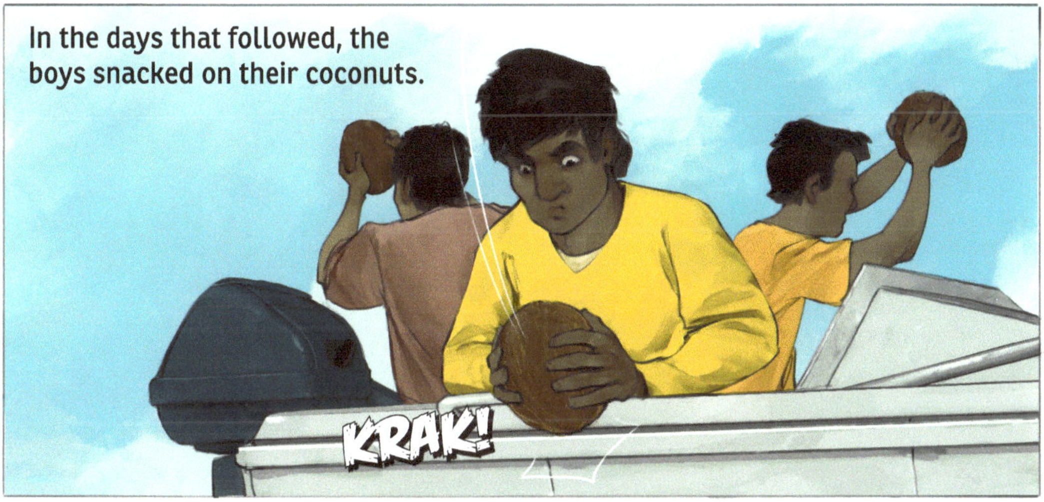
In the days that followed, the boys snacked on their coconuts.
KRAK!

The coconuts gave the boys food and liquids.

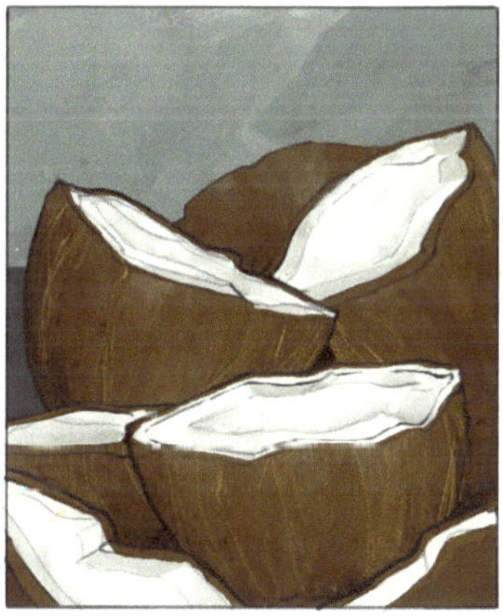

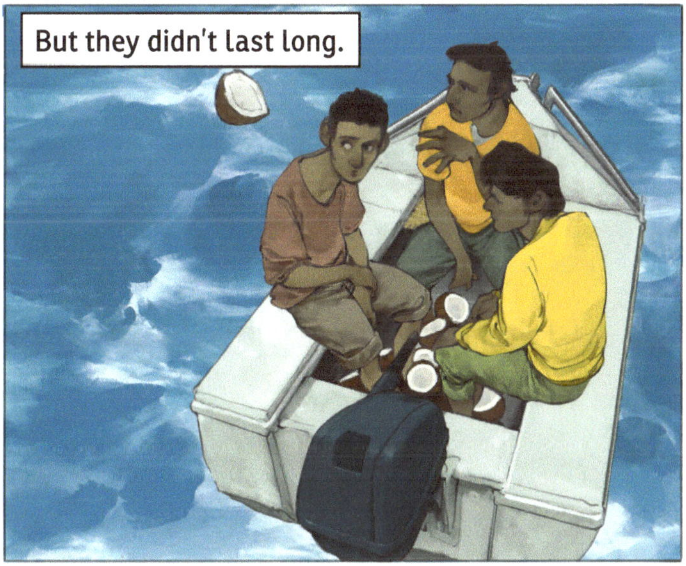
But they didn't last long.

Things were quickly growing dire.

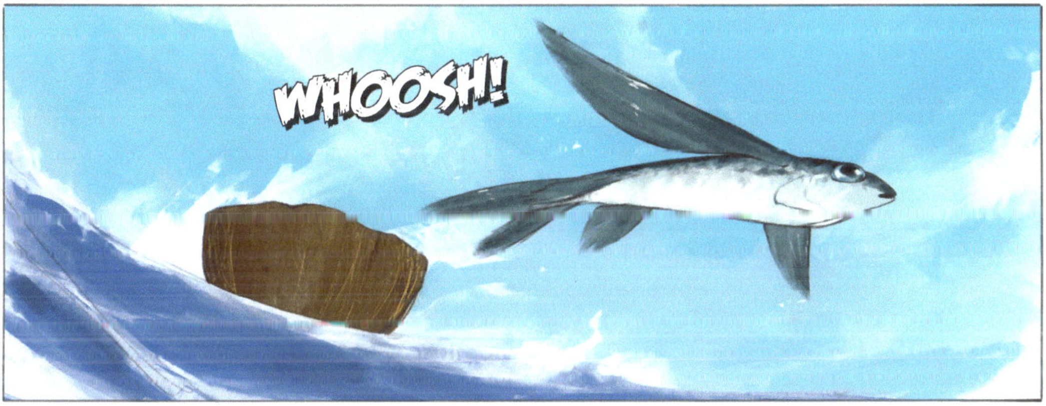
WHOOSH!

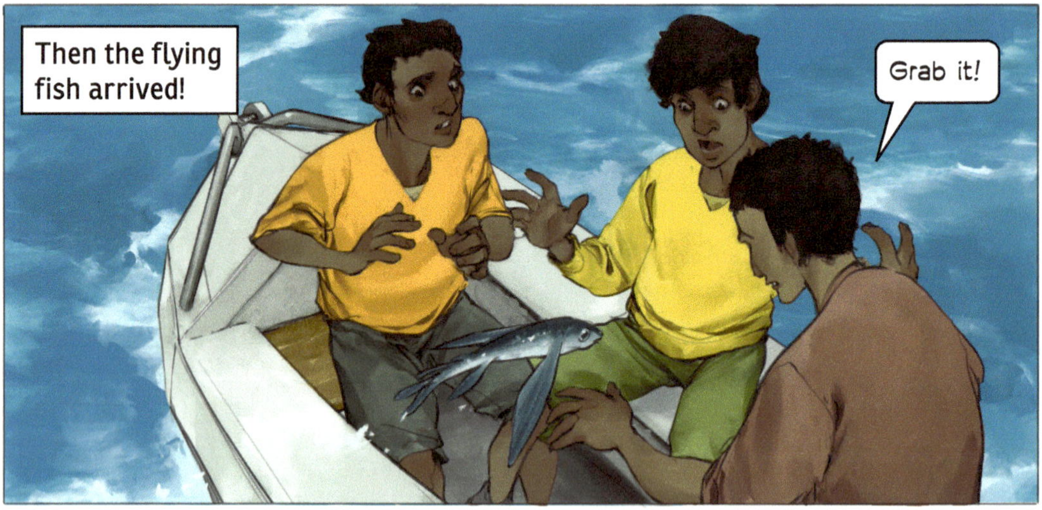
Then the flying fish arrived!
Grab it!

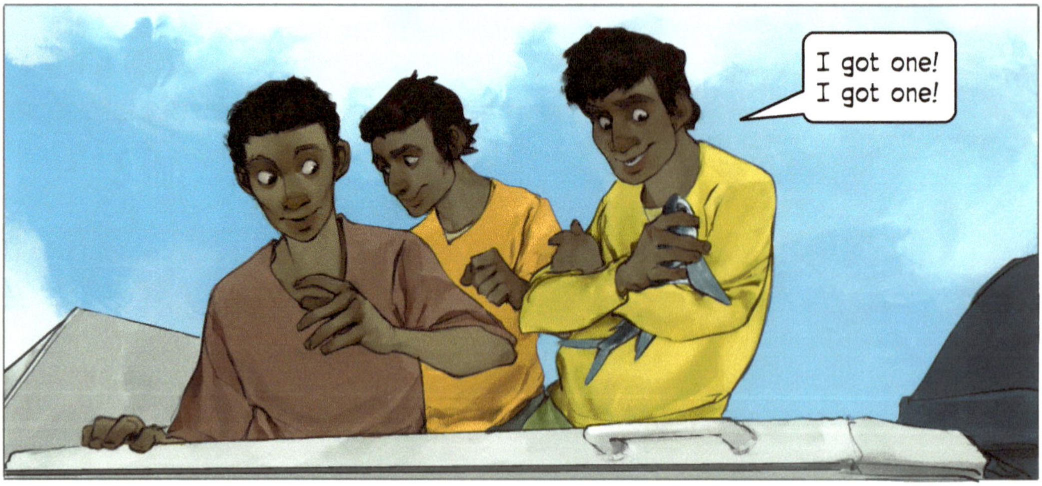
I got one! I got one!

Between the flying fish and a few storms . . .

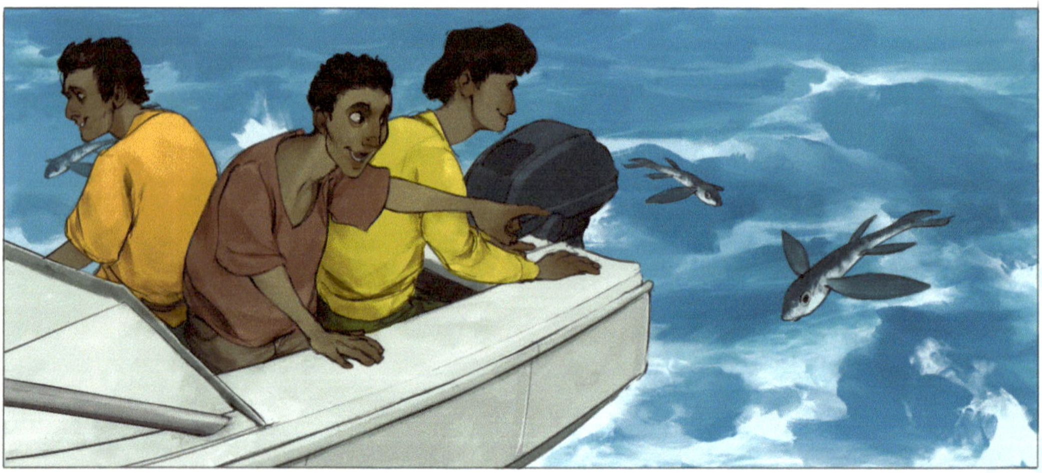

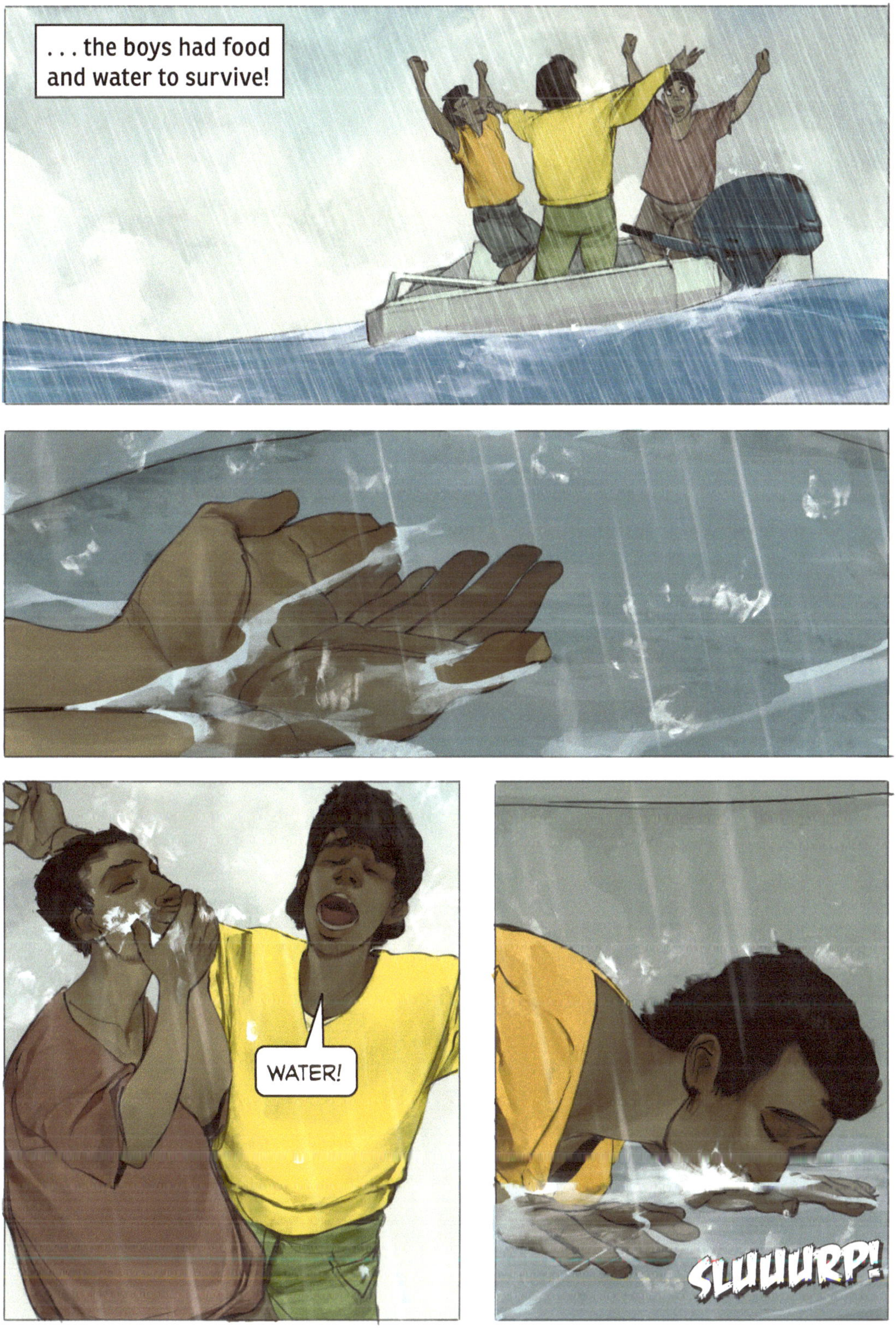

. . . the boys had food and water to survive!
WATER!
SLUUURP!

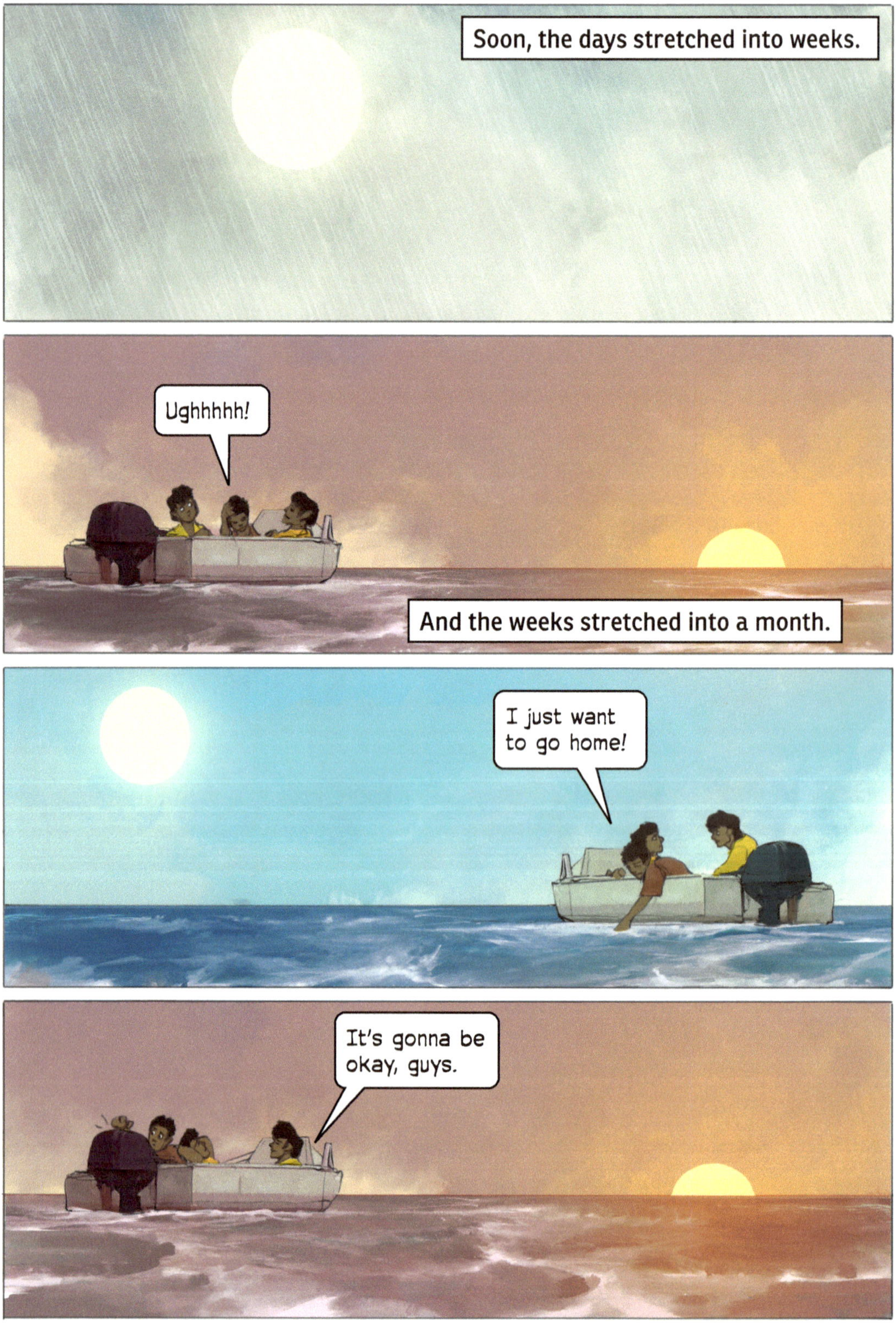

Soon, the days stretched into weeks.
Ughhhhh!
And the weeks stretched into a month.
I just want to go home!
It's gonna be okay, guys.

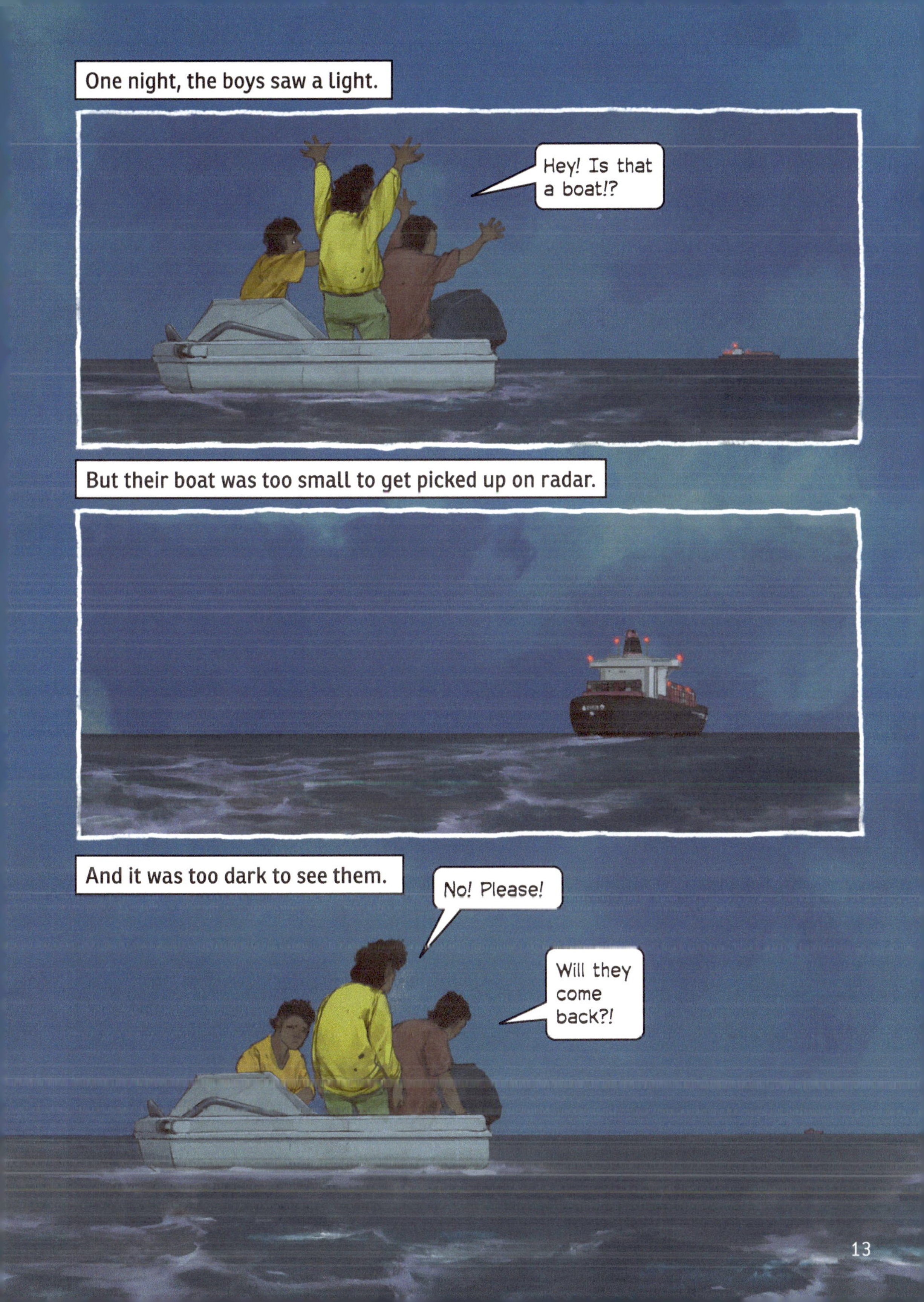
One night, the boys saw a light.
Hey! Is that a boat!?
But their boat was too small to get picked up on radar.
And it was too dark to see them.
No! Please!
Will they come back?!

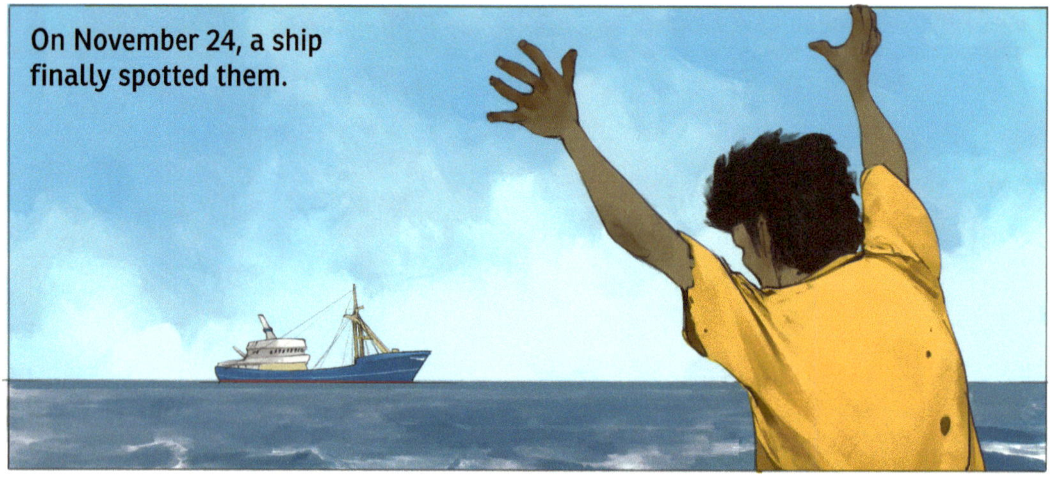
On November 24, a ship finally spotted them.

It had strayed into the same waters by pure luck.

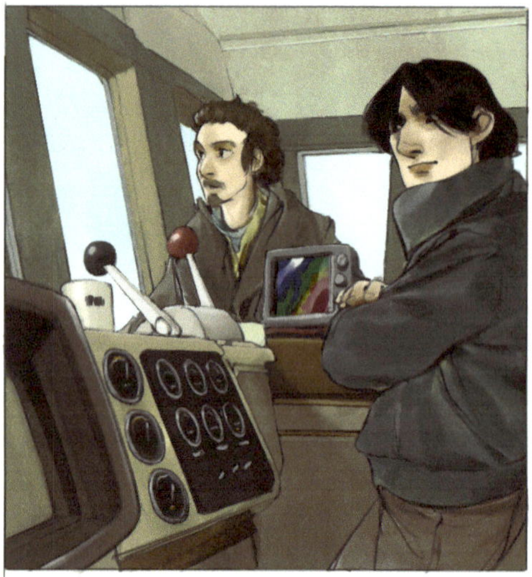

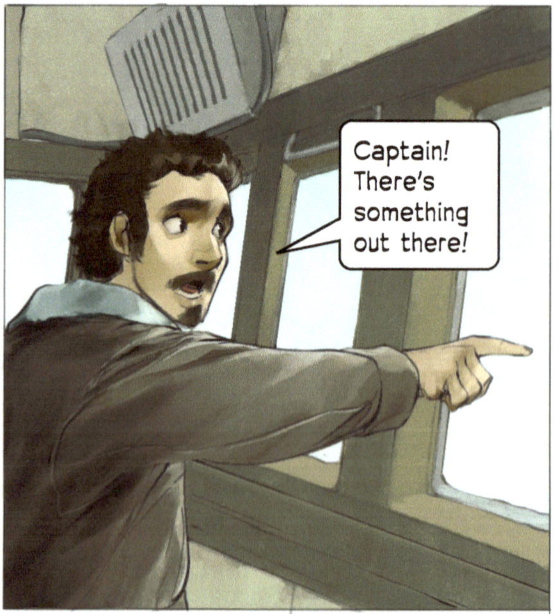
Captain! There's something out there!

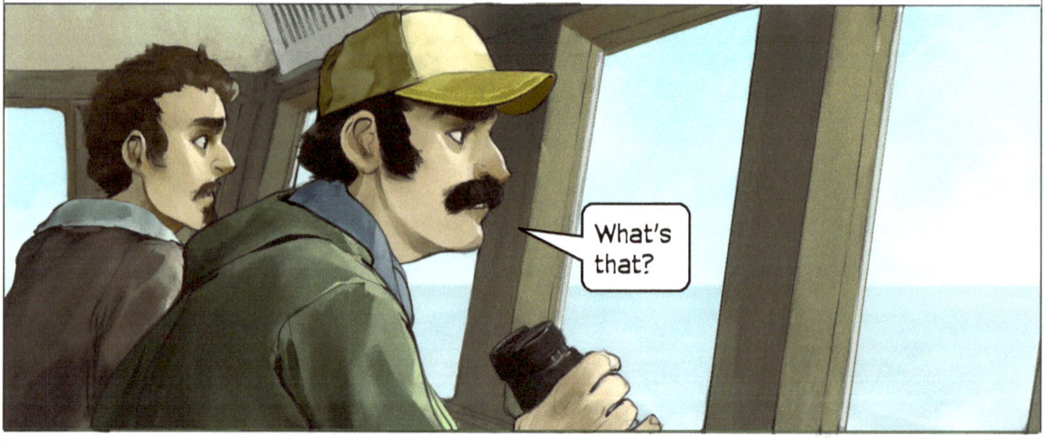
What's that?

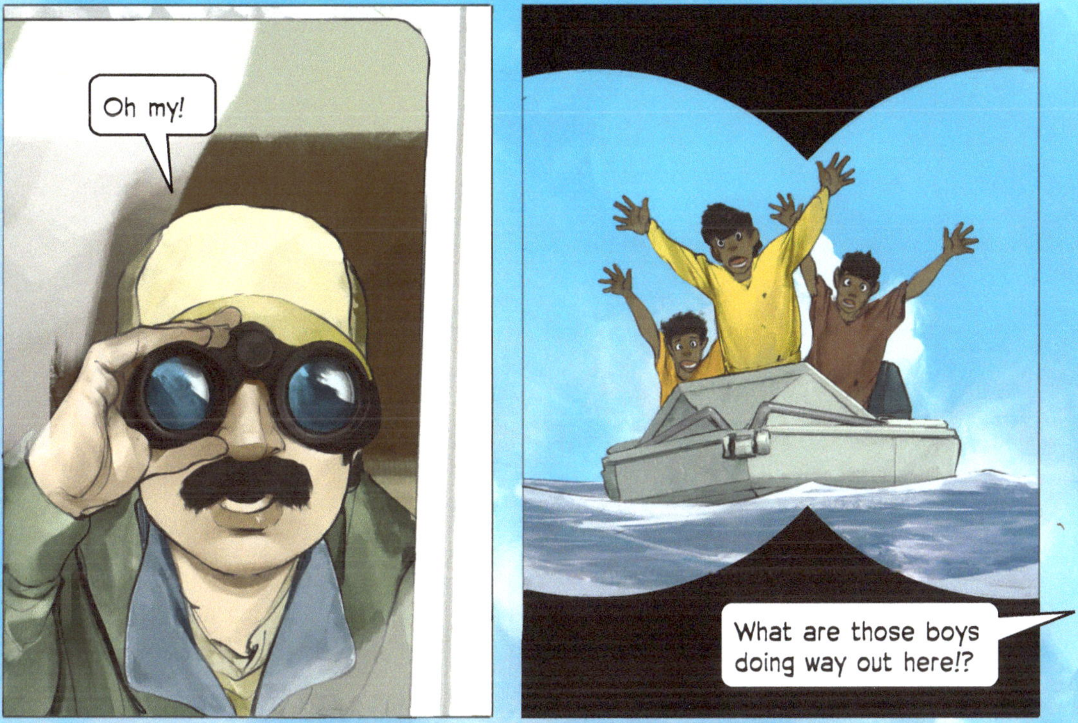
Oh my!
What are those boys doing way out here!?

After more than 50 days lost at sea, the boys were finally rescued.

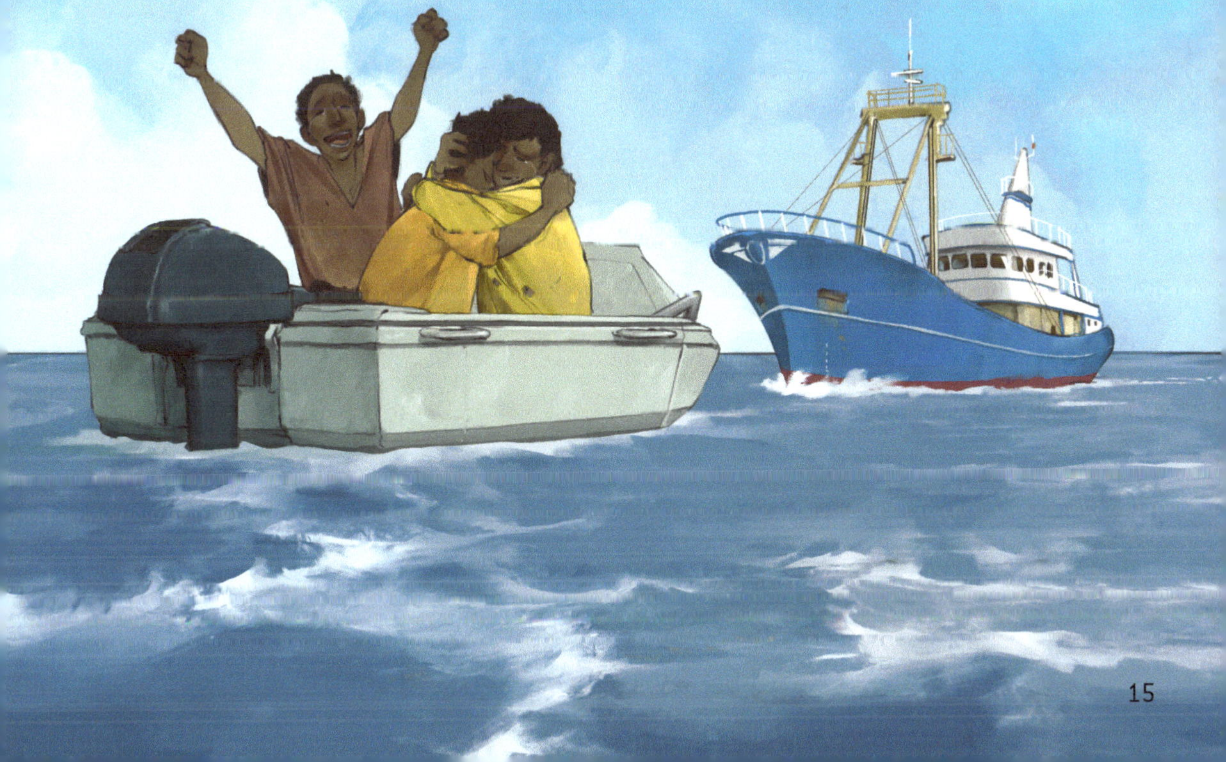

CHAPTER 2
SO LONG, SOLO!

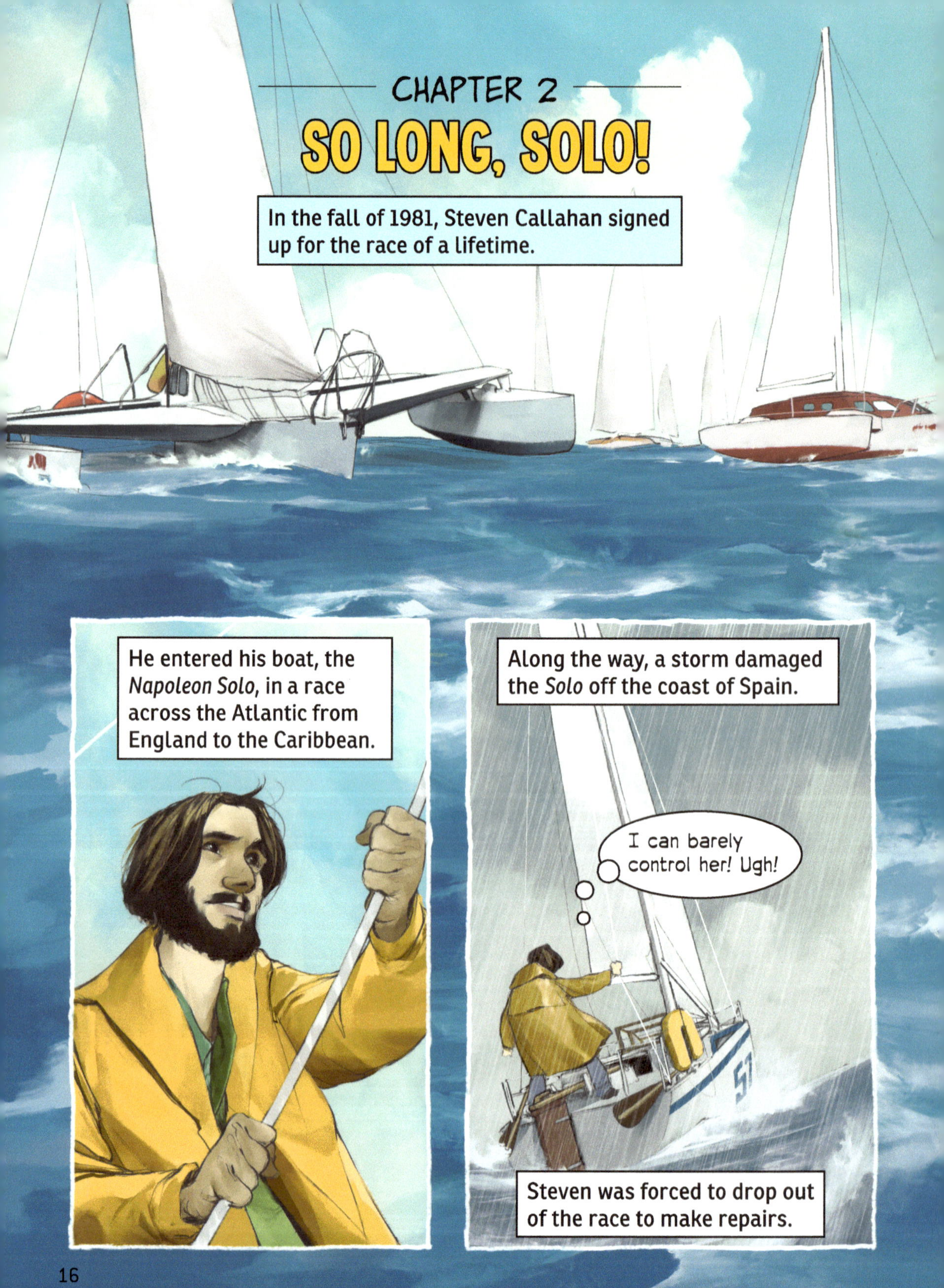
In the fall of 1981, Steven Callahan signed up for the race of a lifetime.
He entered his boat, the *Napoleon Solo*, in a race across the Atlantic from England to the Caribbean.
Along the way, a storm damaged the *Solo* off the coast of Spain.
I can barely control her! Ugh!
Steven was forced to drop out of the race to make repairs.

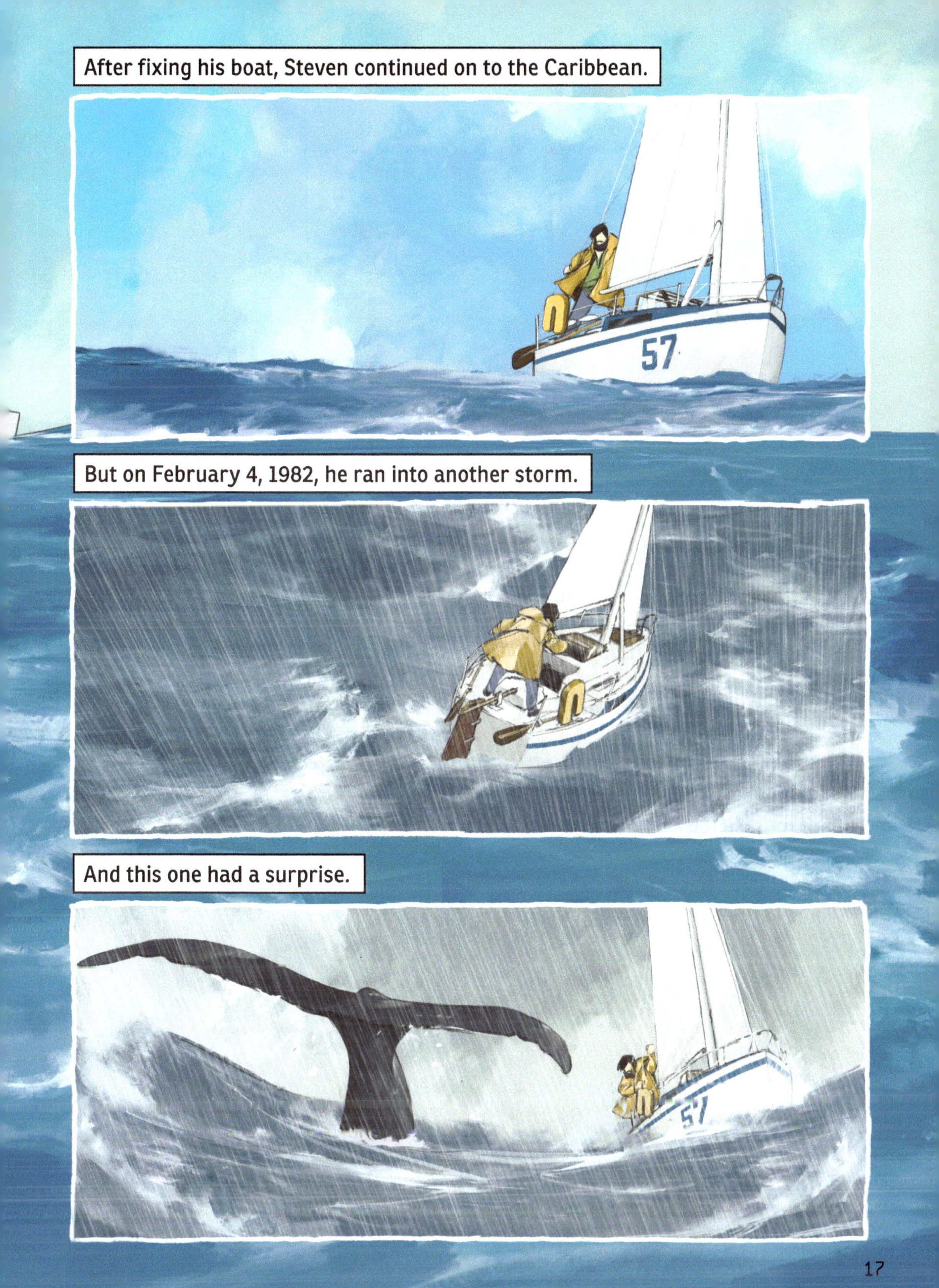
After fixing his boat, Steven continued on to the Caribbean.
57
But on February 4, 1982, he ran into another storm.
And this one had a surprise.
57

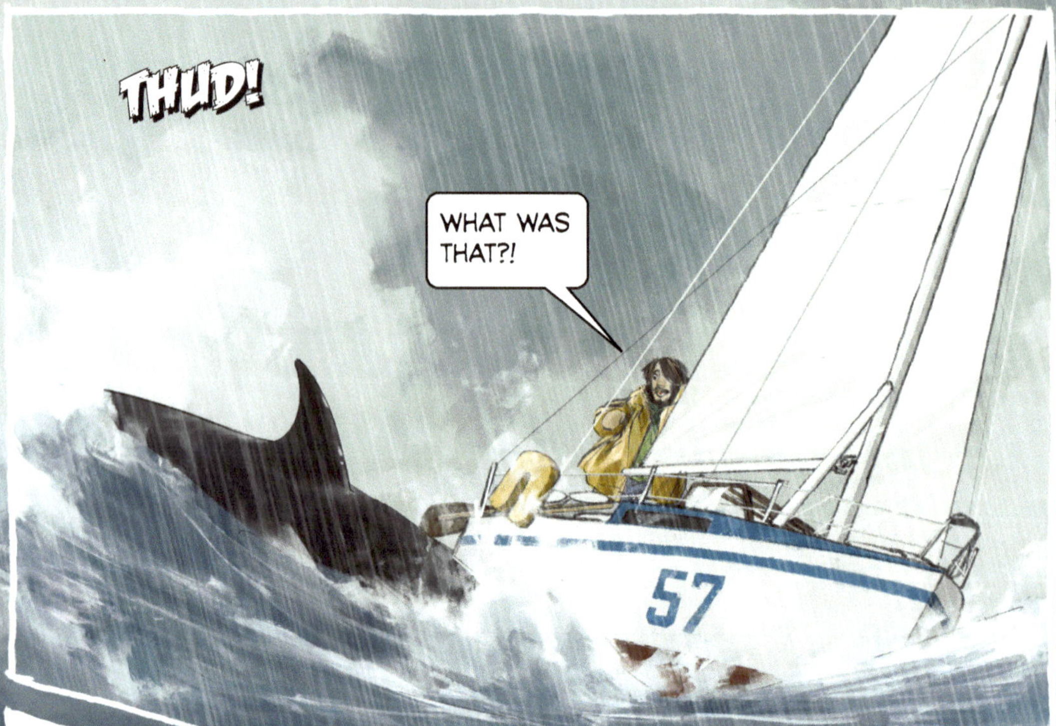
THUD!
WHAT WAS THAT?!
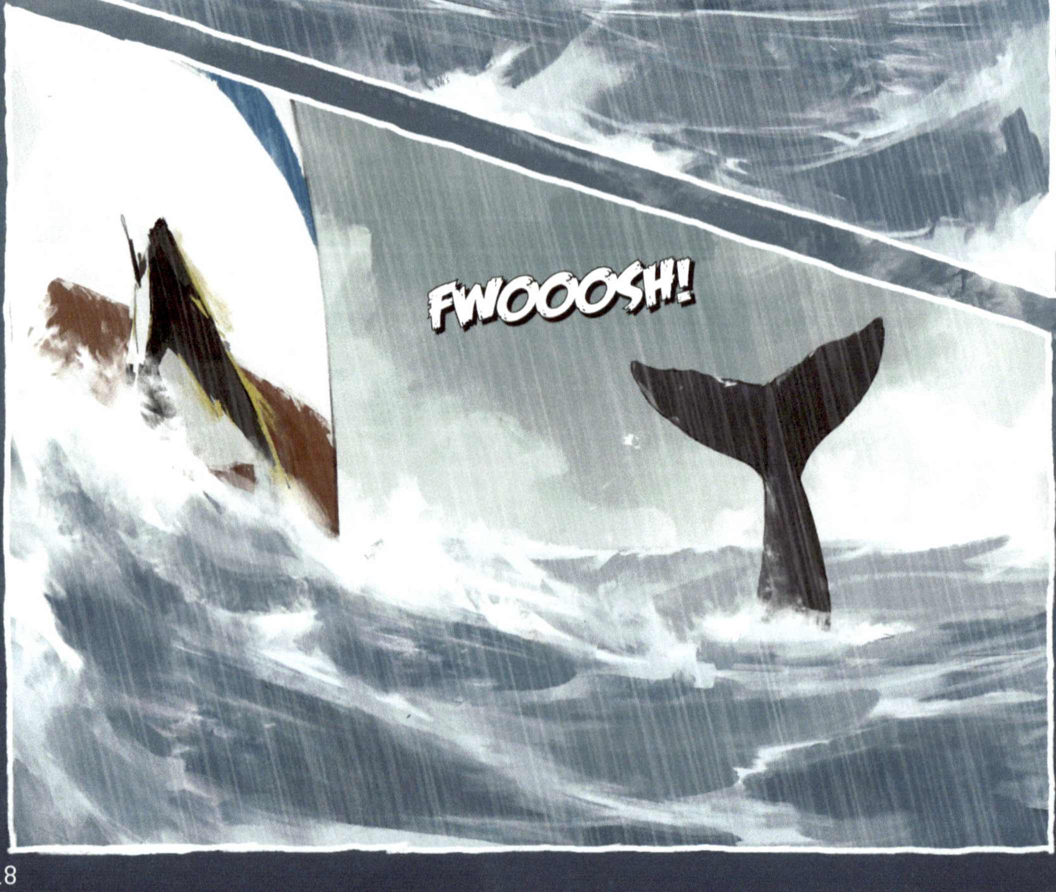
FWOOOSH!

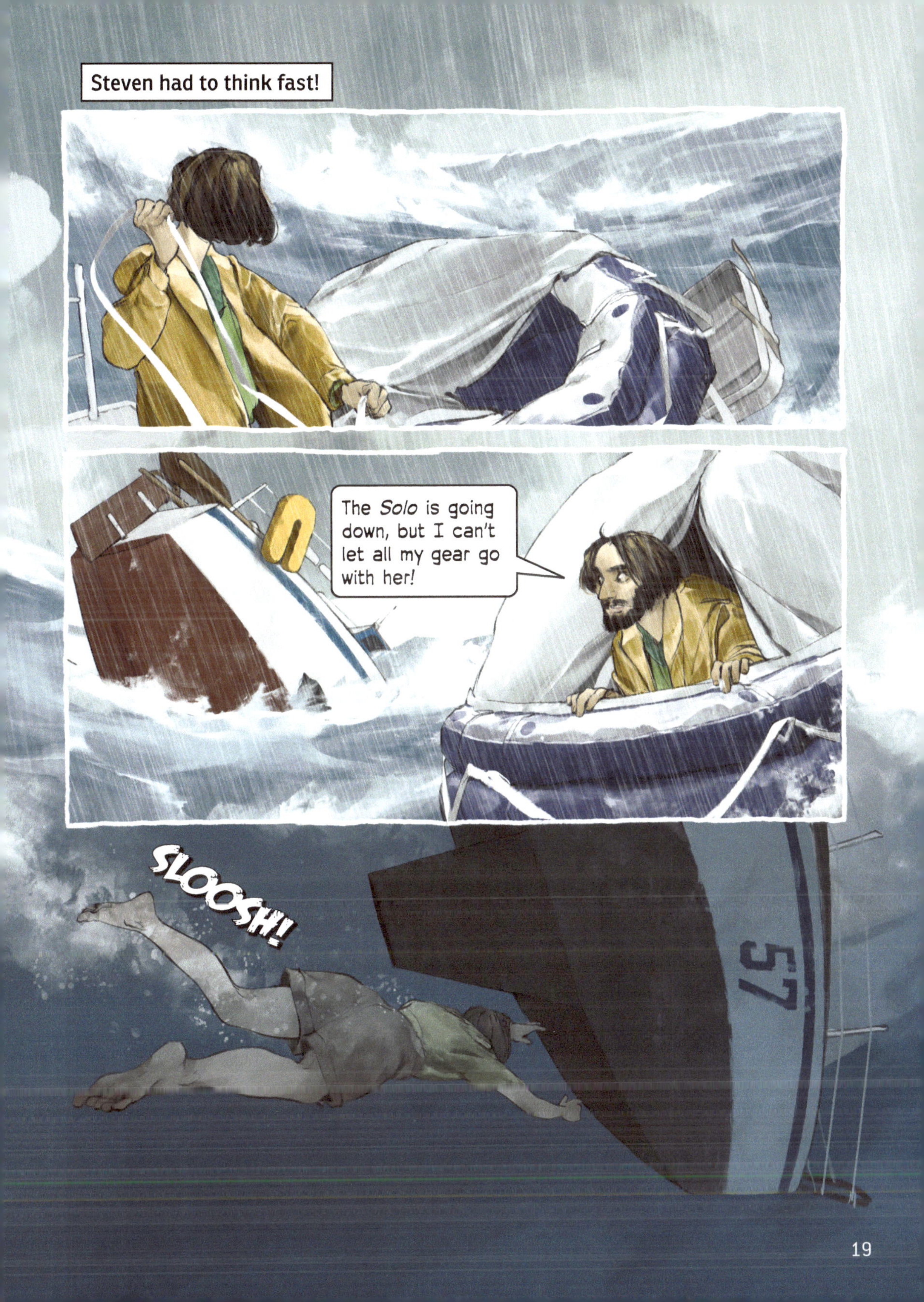

Steven had to think fast!
The *Solo* is going down, but I can't let all my gear go with her!
SLOOSH!
57

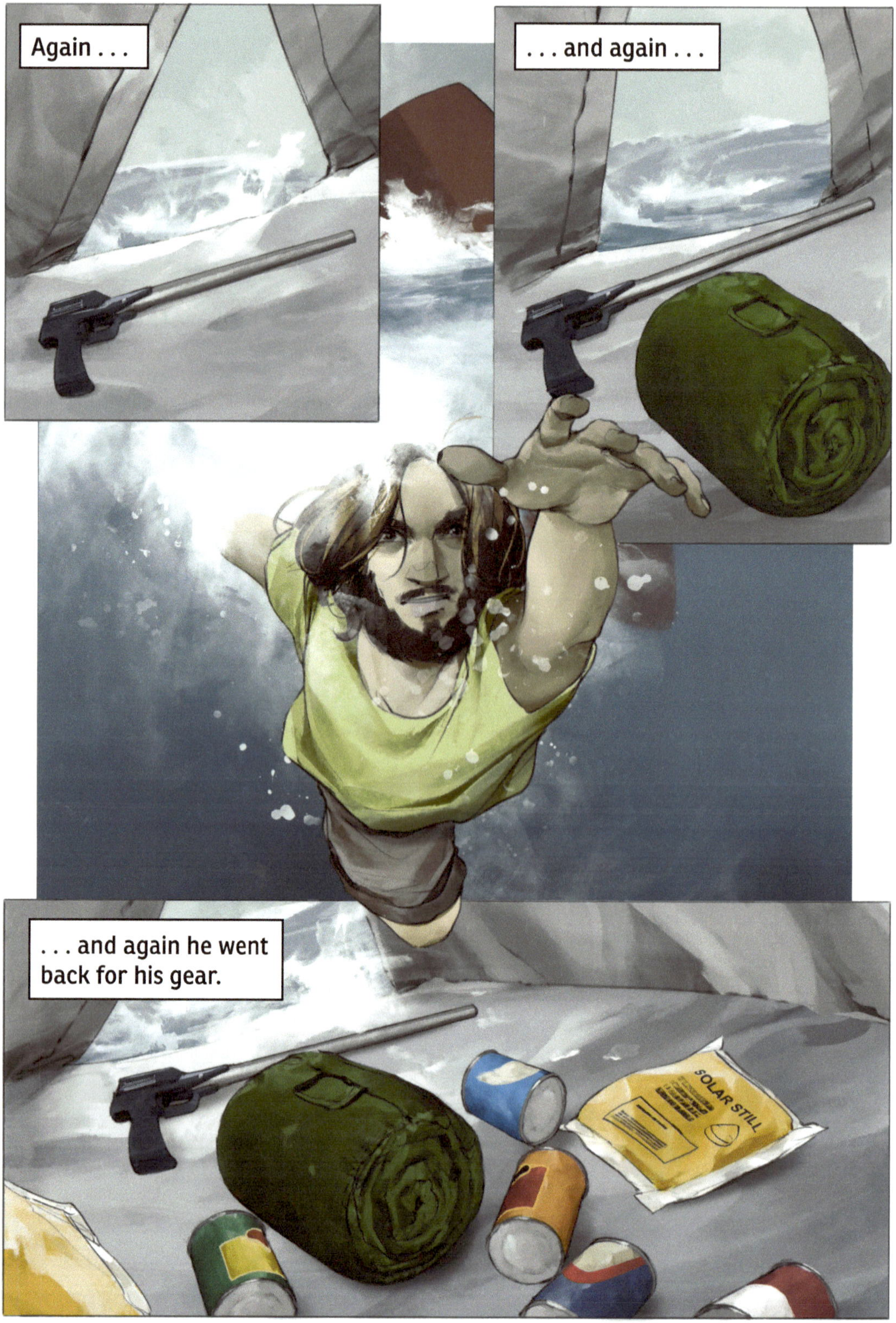
Again . . .
. . . and again . . .
. . . and again he went back for his gear.
SOLAR STILL

All in all, he was able to grab a lot of helpful items!

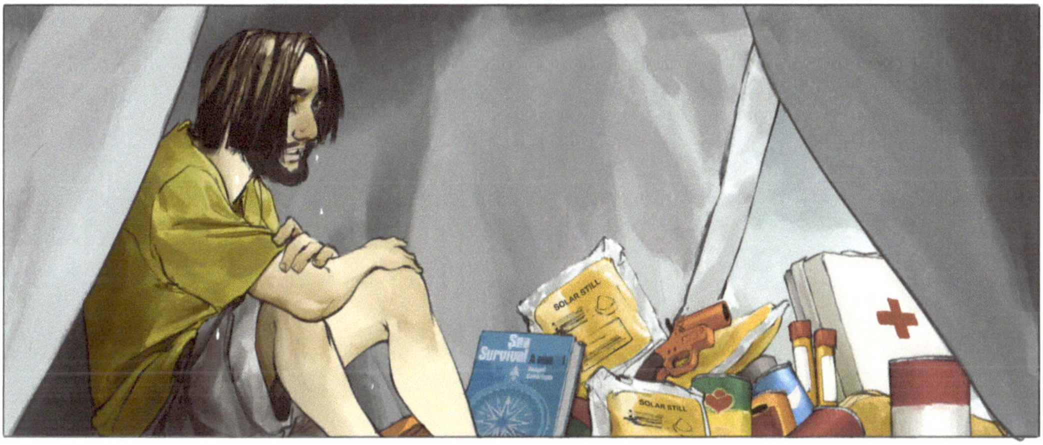
SOLAR STILL
SOLAR STILL

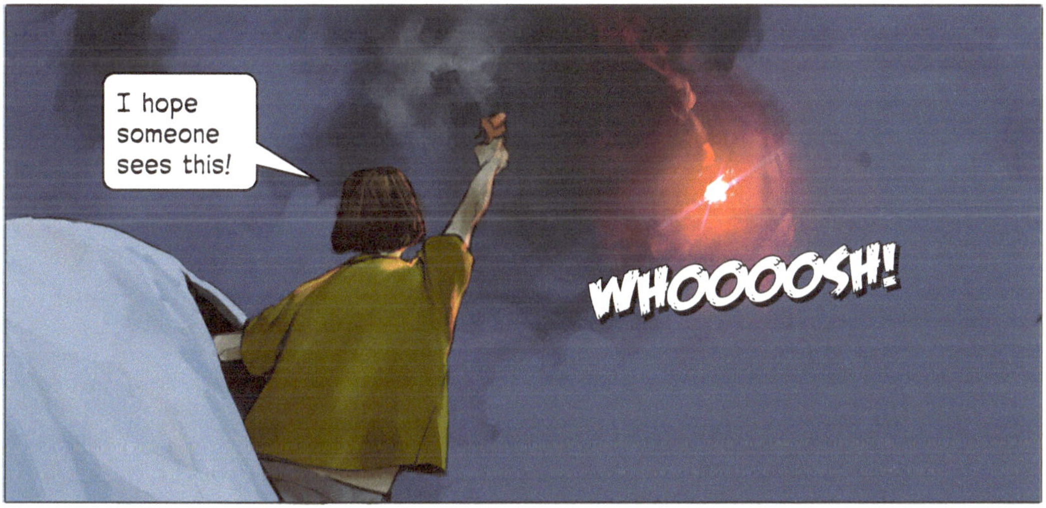
I hope someone sees this!
WHOOOOOSH!

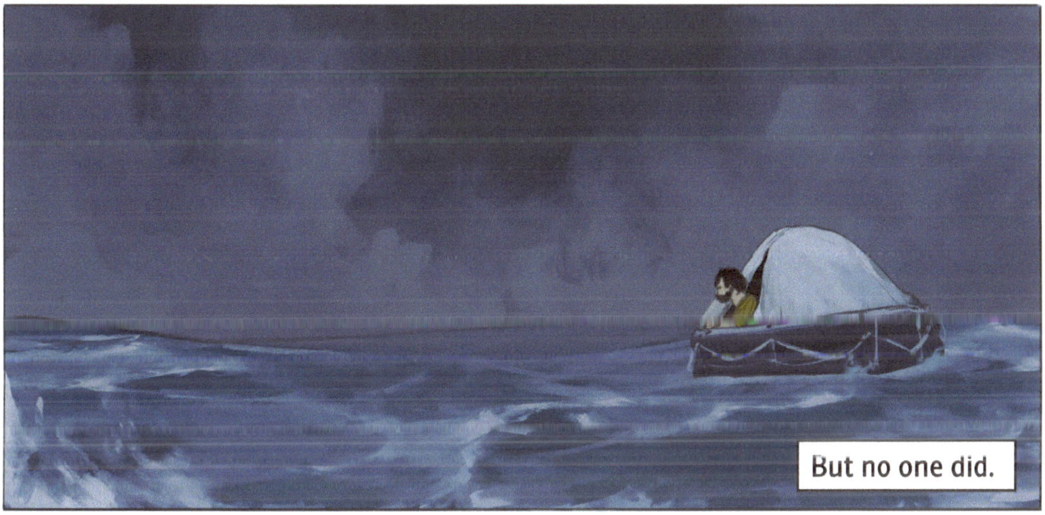
But no one did.

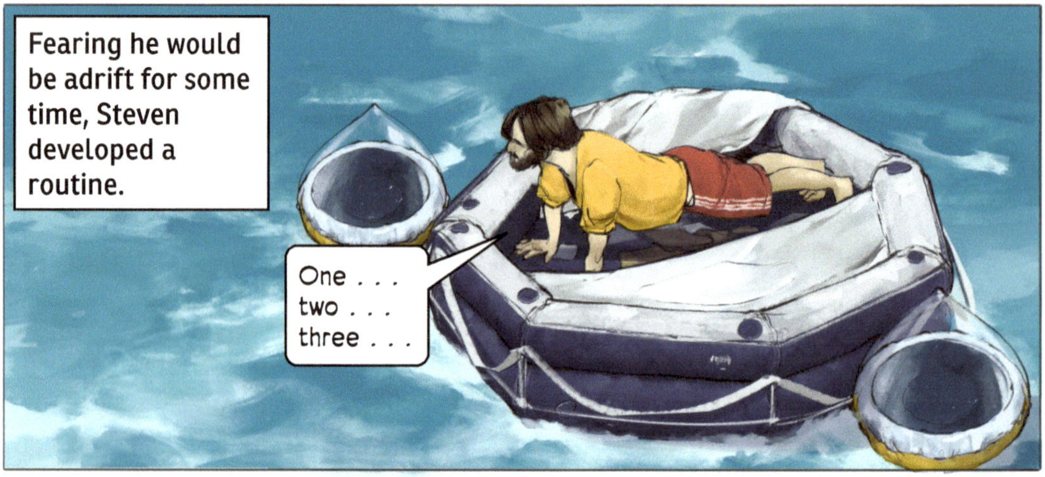
Fearing he would be adrift for some time, Steven developed a routine.
One . . . two . . . three . . .

He built ways to collect rainwater, and he fished regularly.

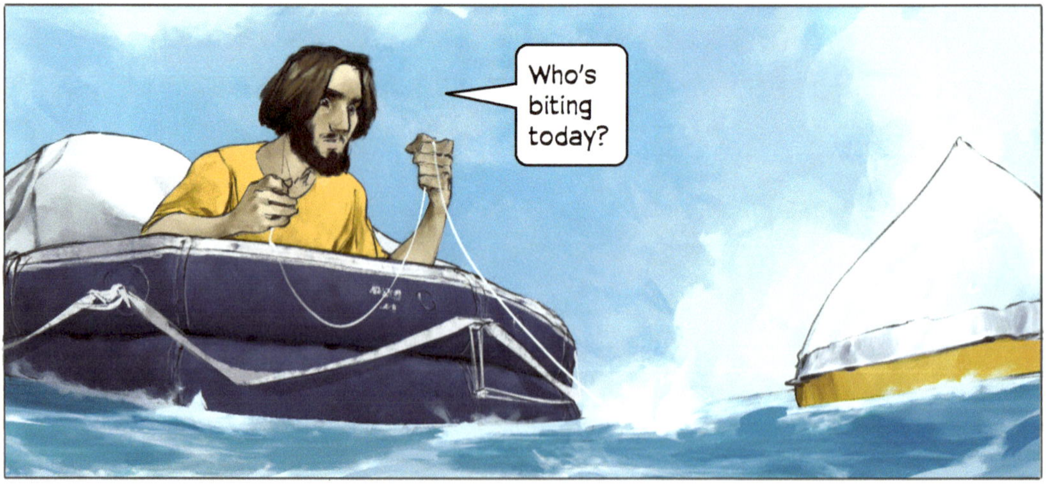
Who's biting today?

He even tried to guide his raft to safety.

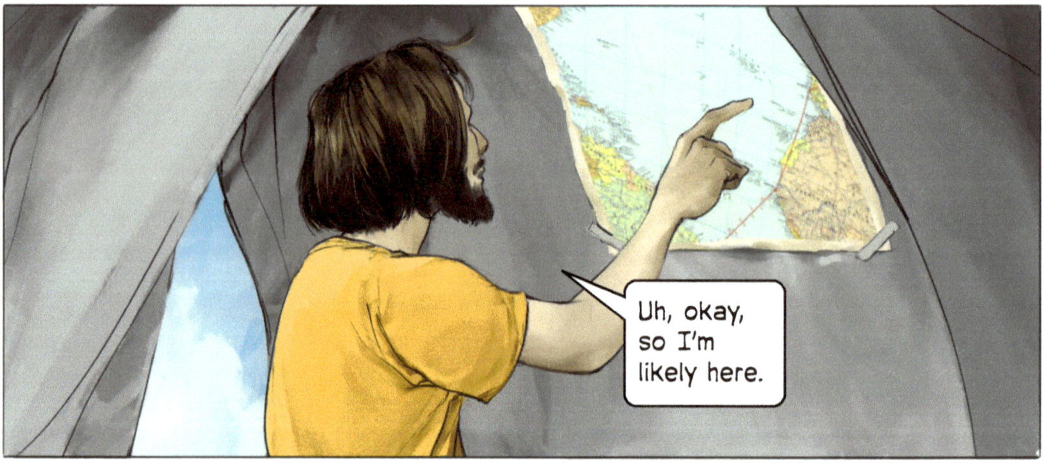
Uh, okay, so I'm likely here.

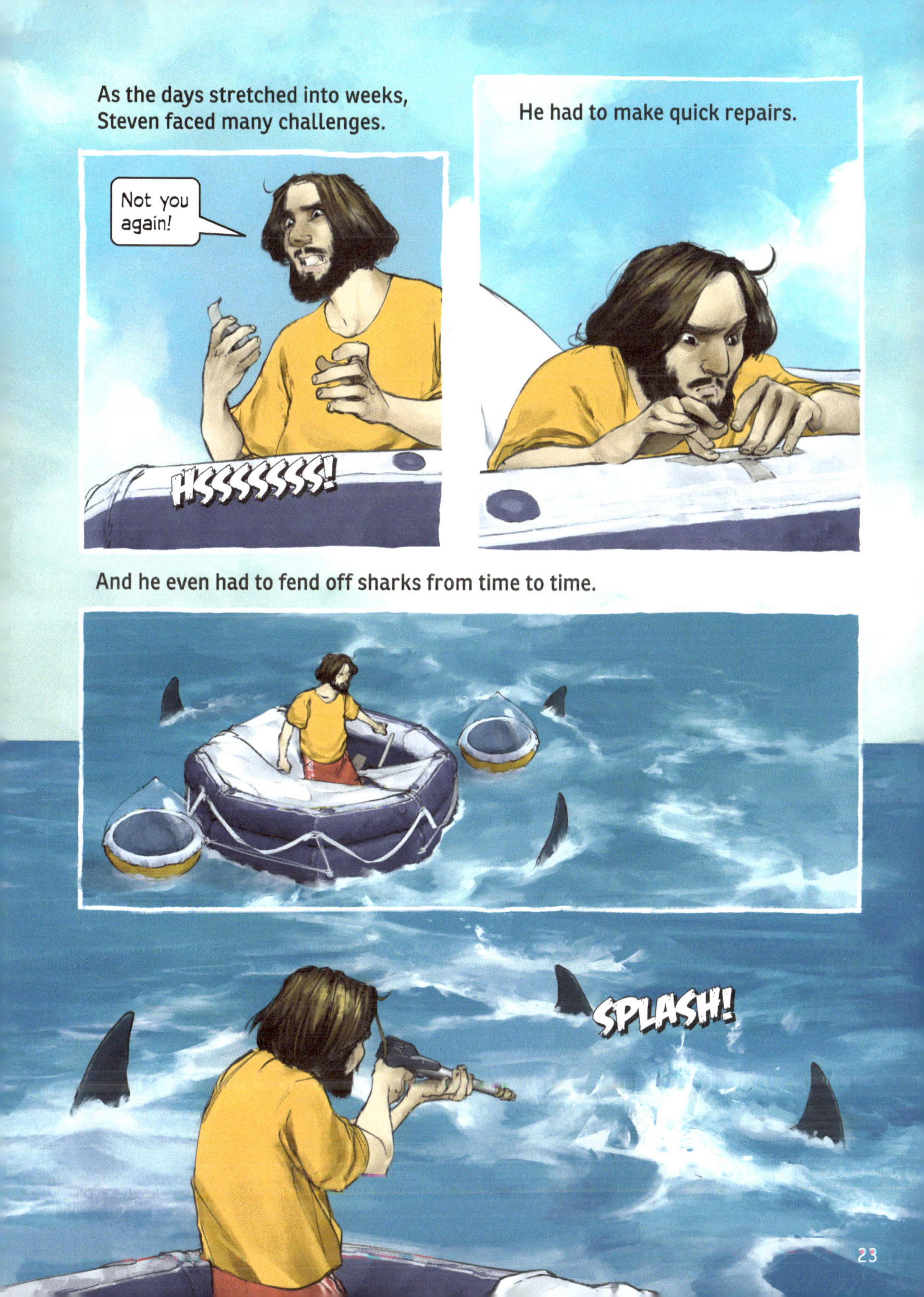
As the days stretched into weeks, Steven faced many challenges.
Not you again!
HSSSSSSS!
He had to make quick repairs.
And he even had to fend off sharks from time to time.
SPLASH!

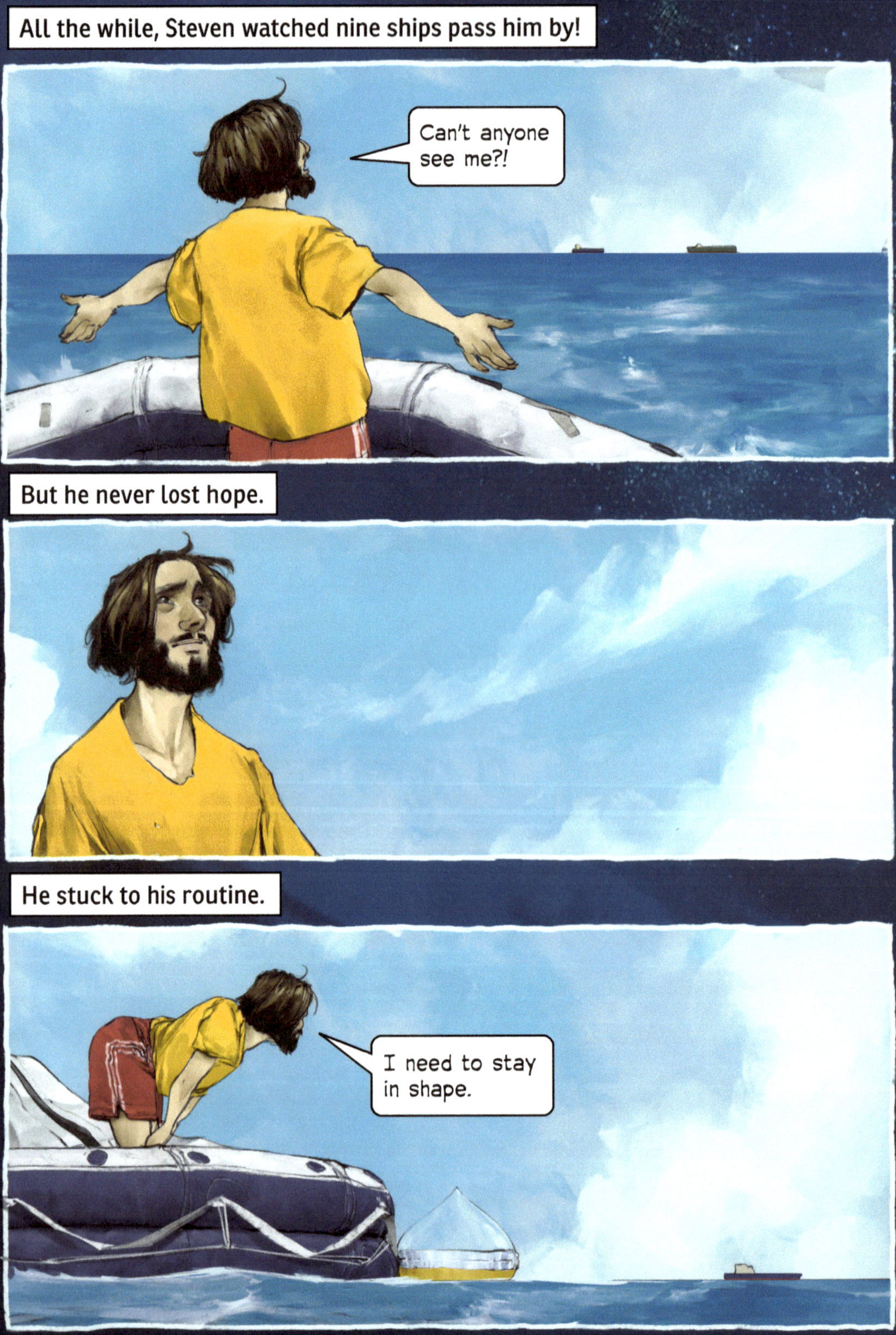
All the while, Steven watched nine ships pass him by!
Can't anyone see me?!
But he never lost hope.
He stuck to his routine.
I need to stay in shape.

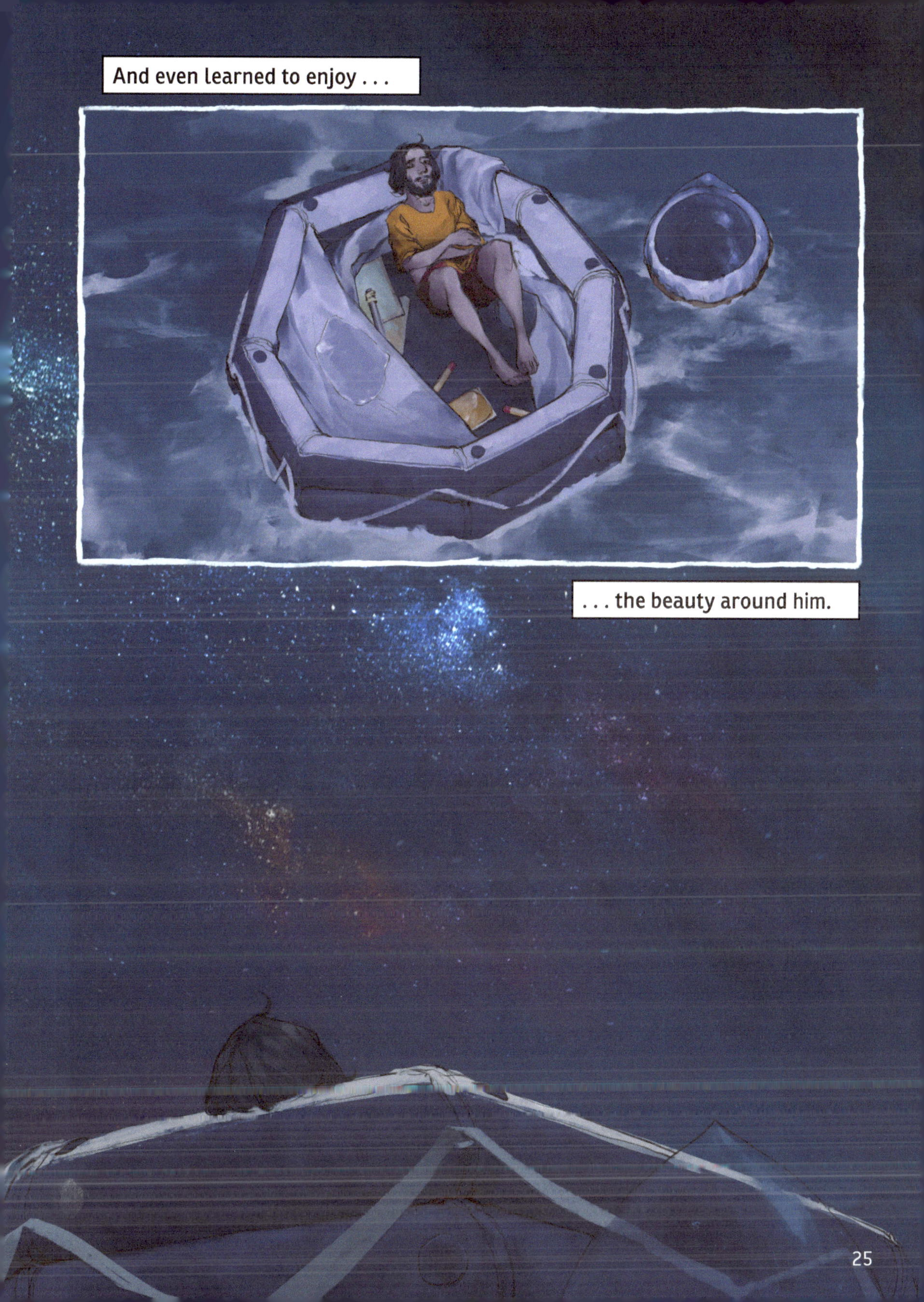
And even learned to enjoy . . .
. . . the beauty around him.

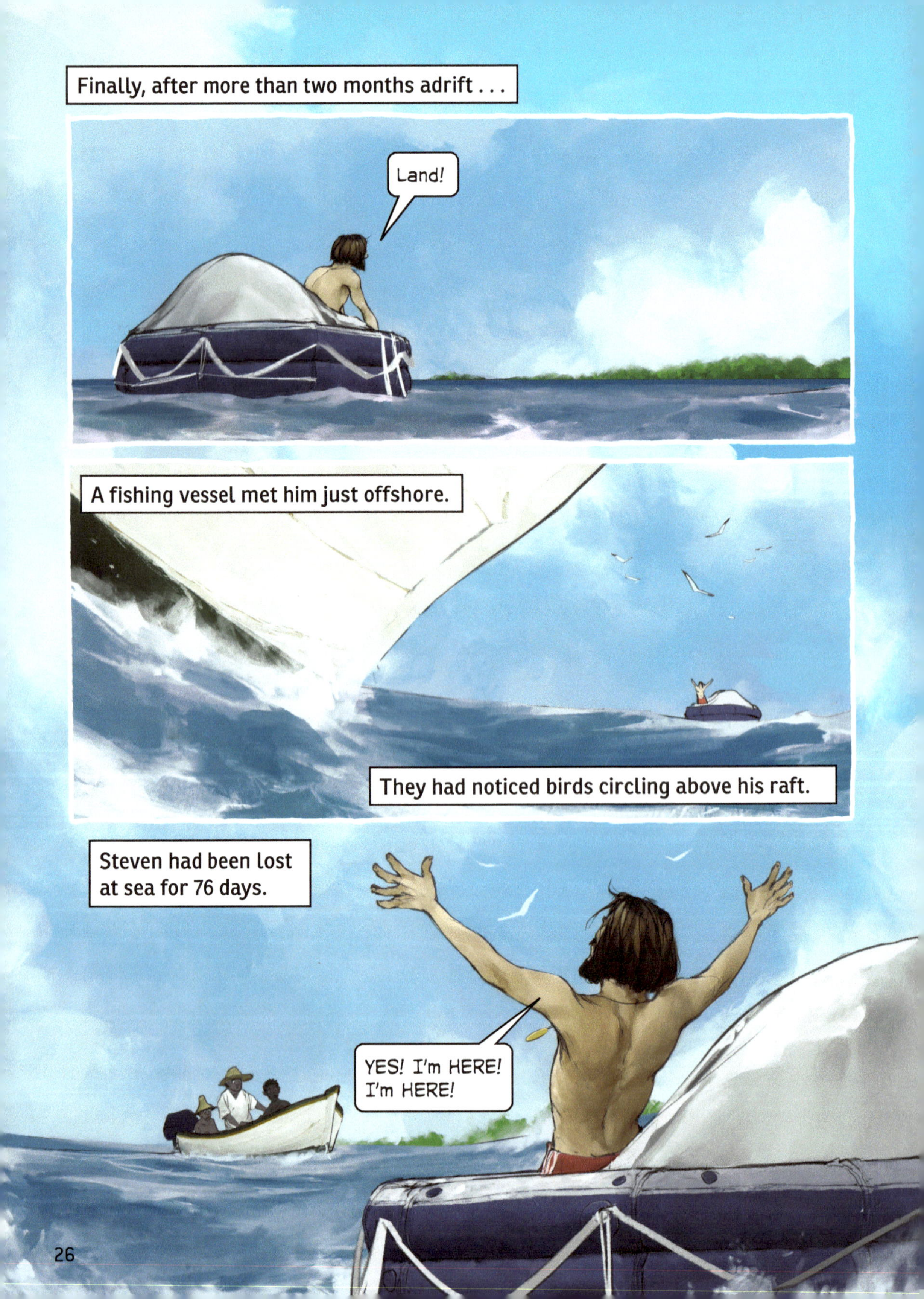
Finally, after more than two months adrift . . .
Land!
A fishing vessel met him just offshore.
They had noticed birds circling above his raft.
Steven had been lost at sea for 76 days.
YES! I'm HERE! I'm HERE!

He was rescued on the island of Marie-Galante, in the Caribbean.

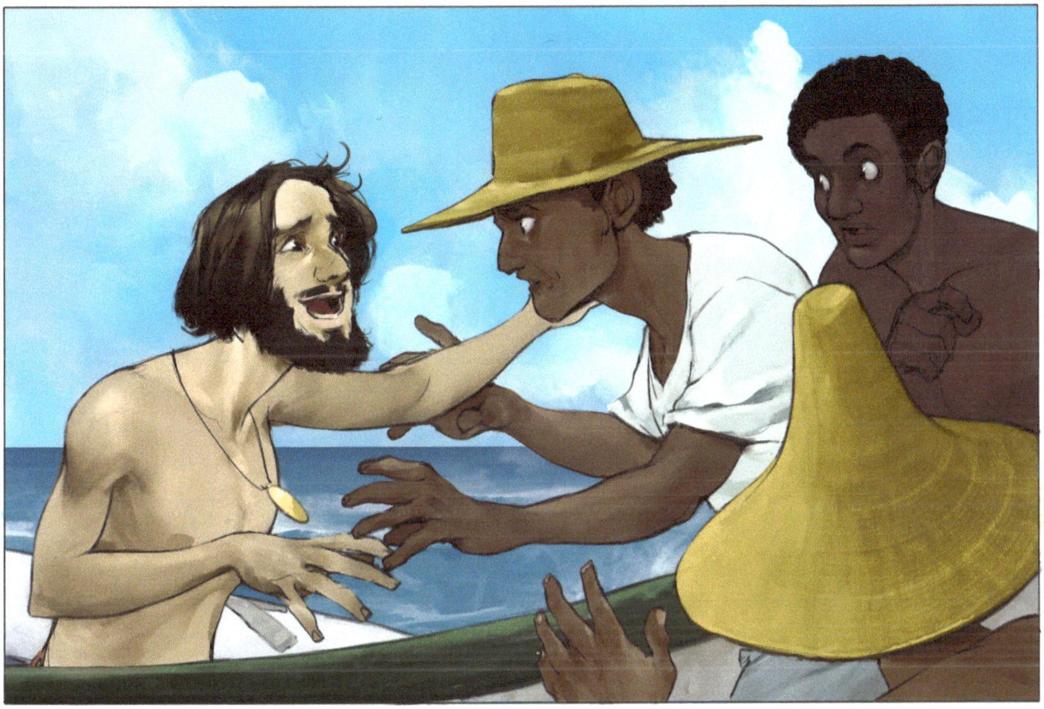

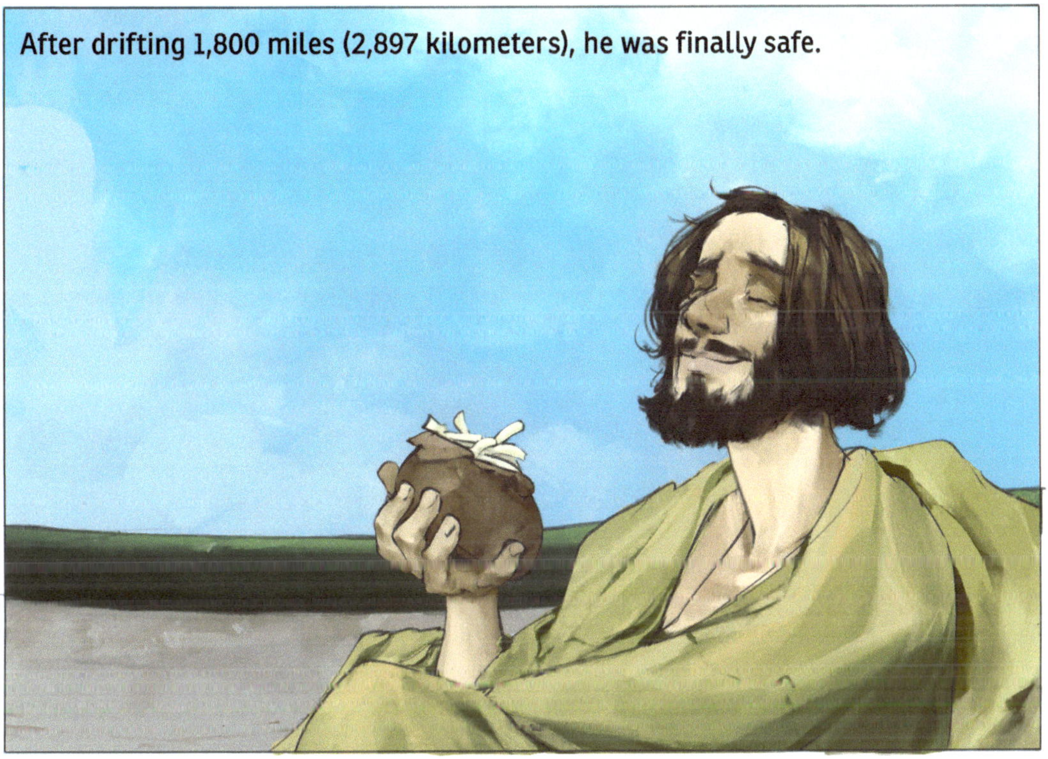
After drifting 1,800 miles (2,897 kilometers), he was finally safe.

CHAPTER 3

DISASTER DOWN UNDER!

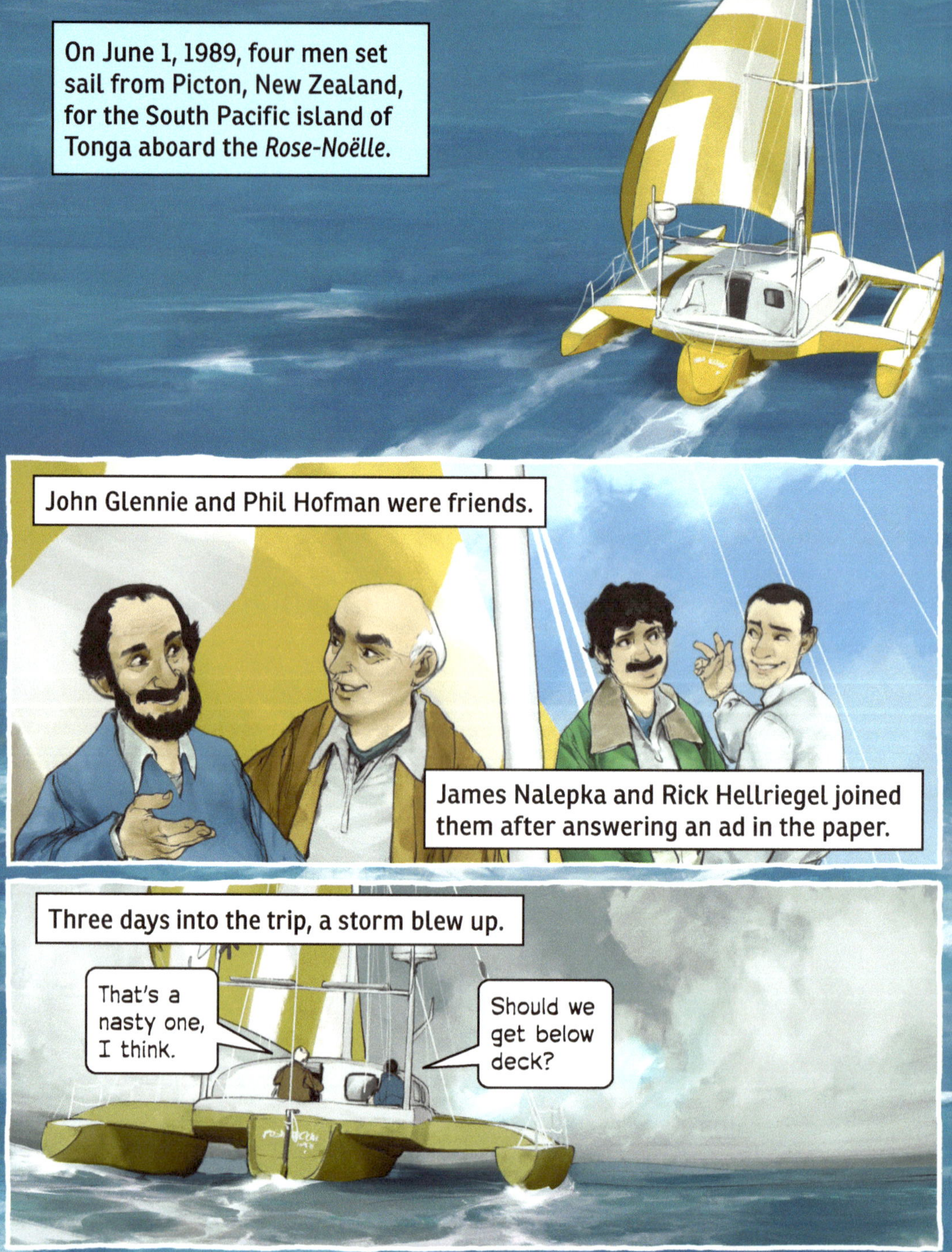
On June 1, 1989, four men set sail from Picton, New Zealand, for the South Pacific island of Tonga aboard the *Rose-Noëlle*.
John Glennie and Phil Hofman were friends.
James Nalepka and Rick Hellriegel joined them after answering an ad in the paper.
Three days into the trip, a storm blew up.
That's a nasty one, I think.
Should we get below deck?

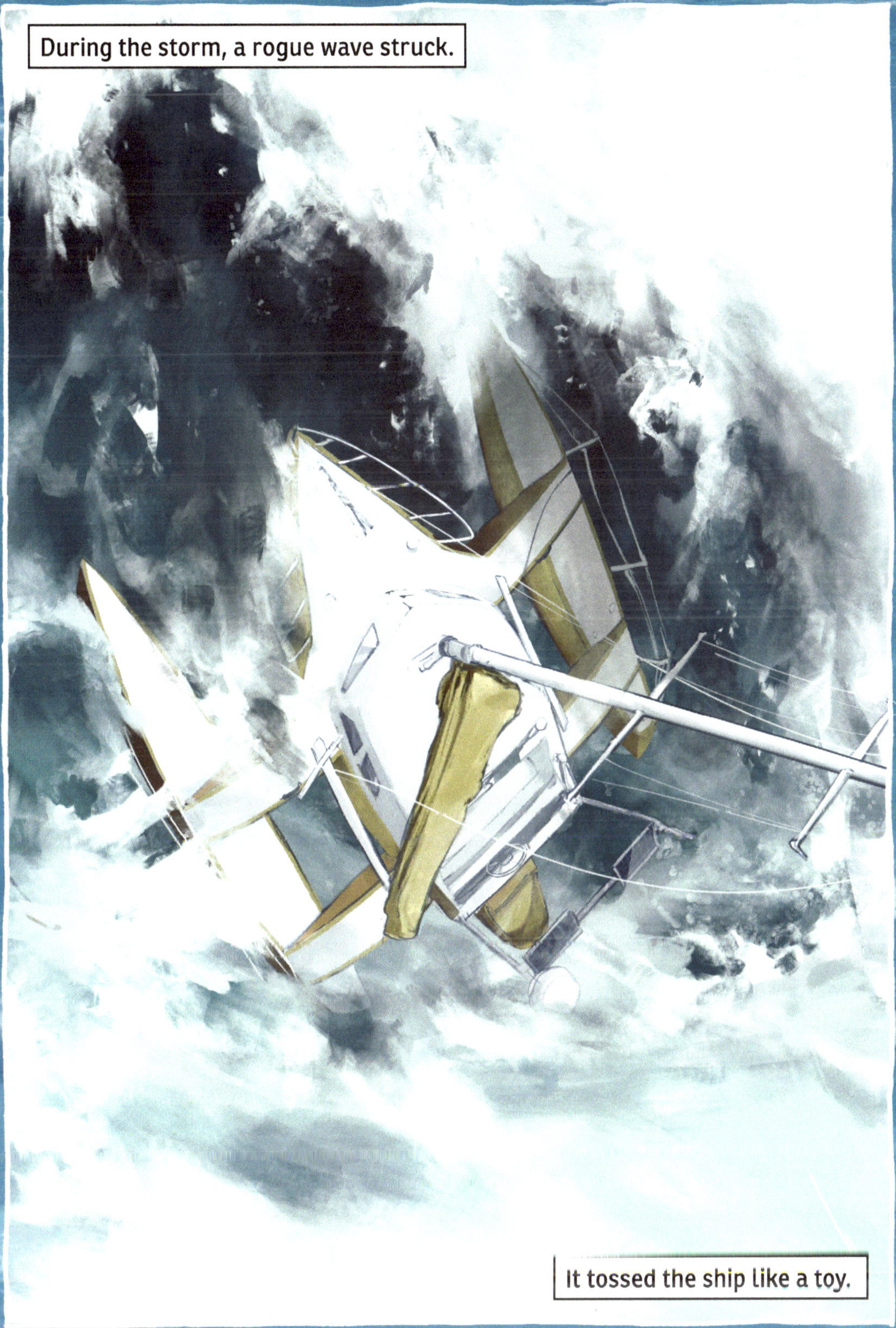
During the storm, a rogue wave struck.
It tossed the ship like a toy.

In an instant, the *Rose-Noëlle* capsized!

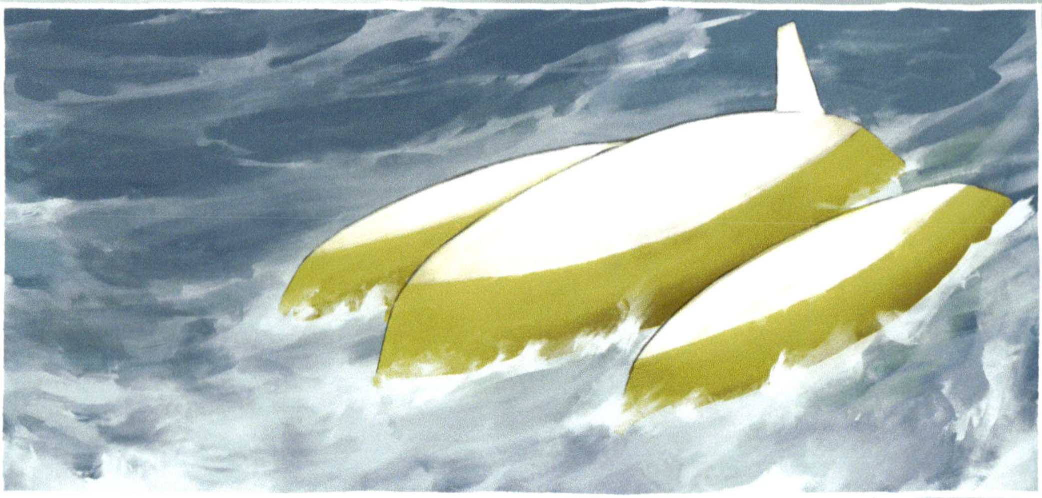

Below deck, the cabin began to flood.

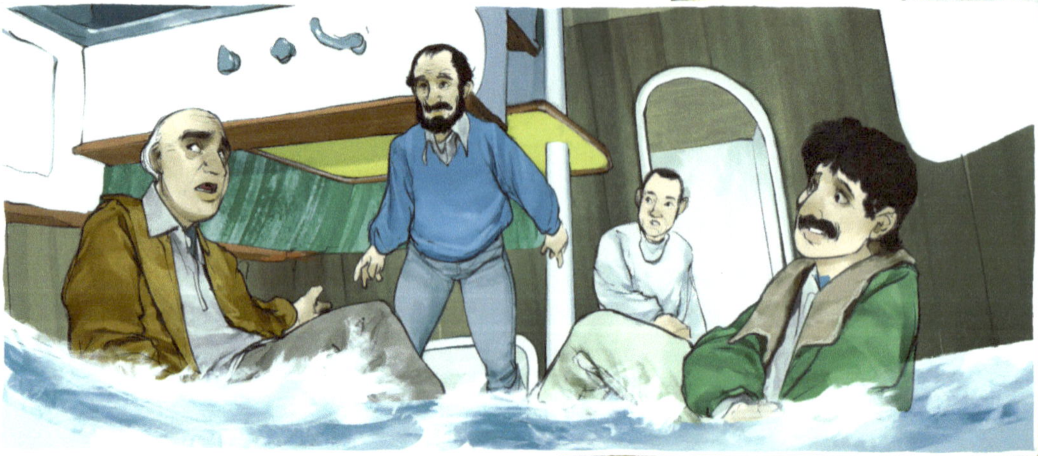

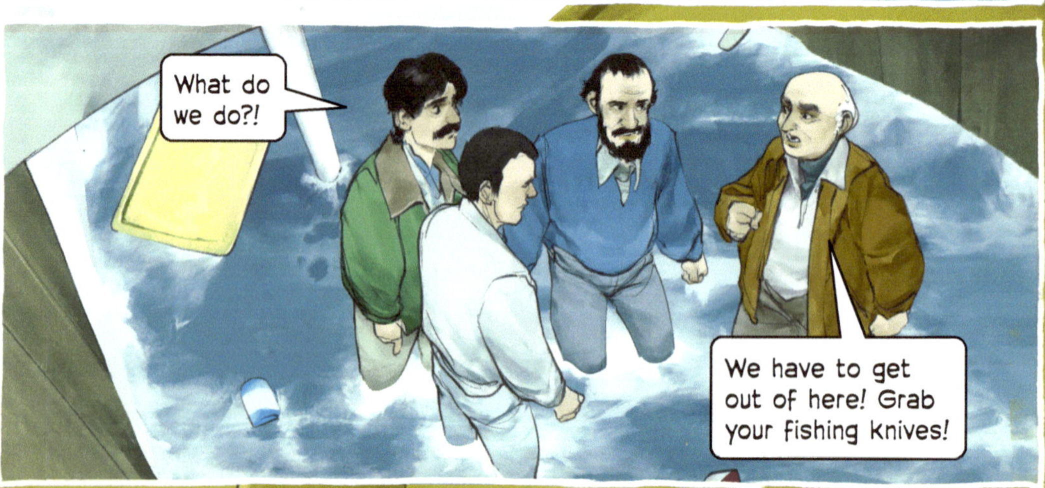
What do we do?!
We have to get out of here! Grab your fishing knives!

Together, they began cutting an escape hatch.

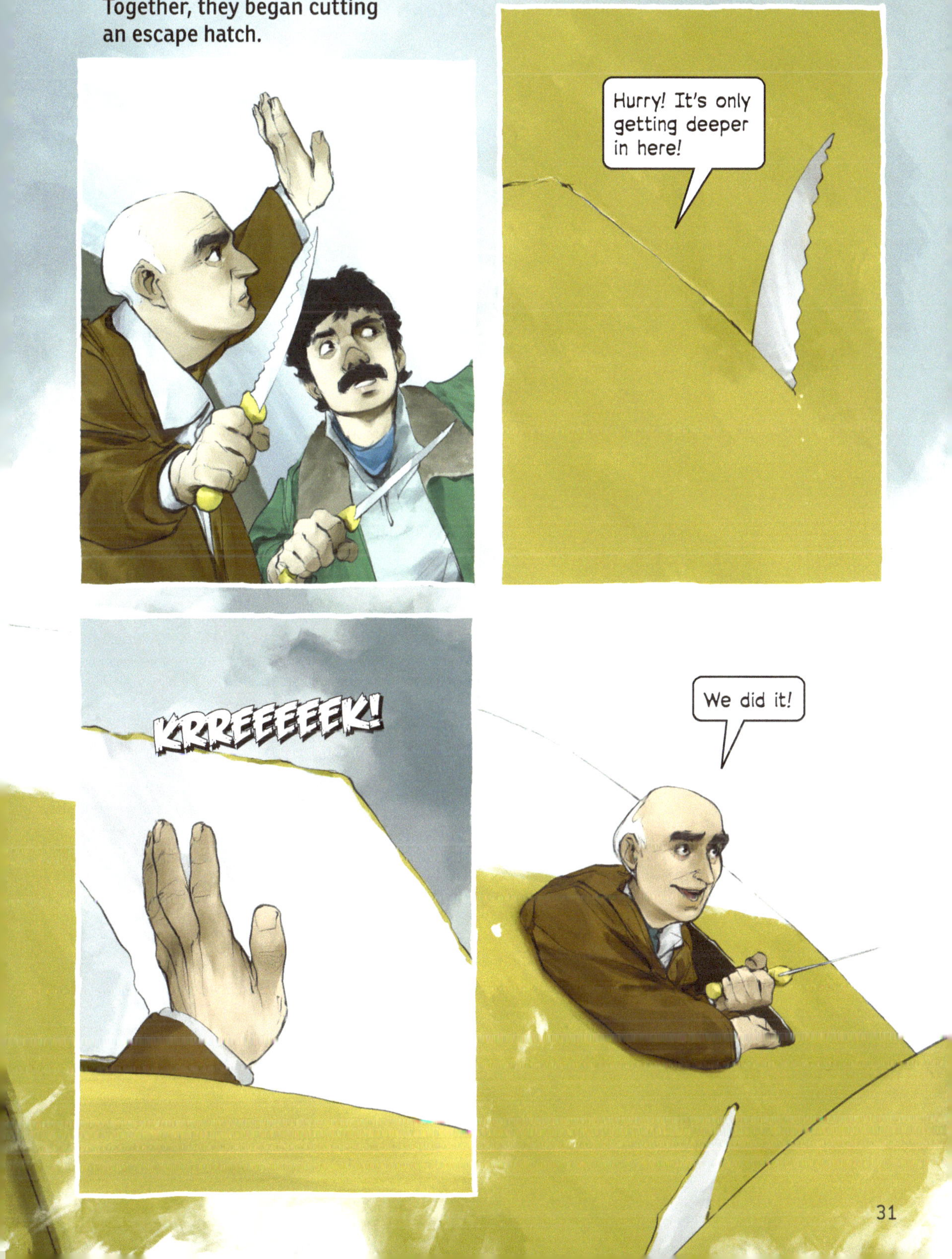
Hurry! It's only getting deeper in here!
KRREEEEEEK!
We did it!

The crew of the *Rose-Noëlle* had survived nature's first test.

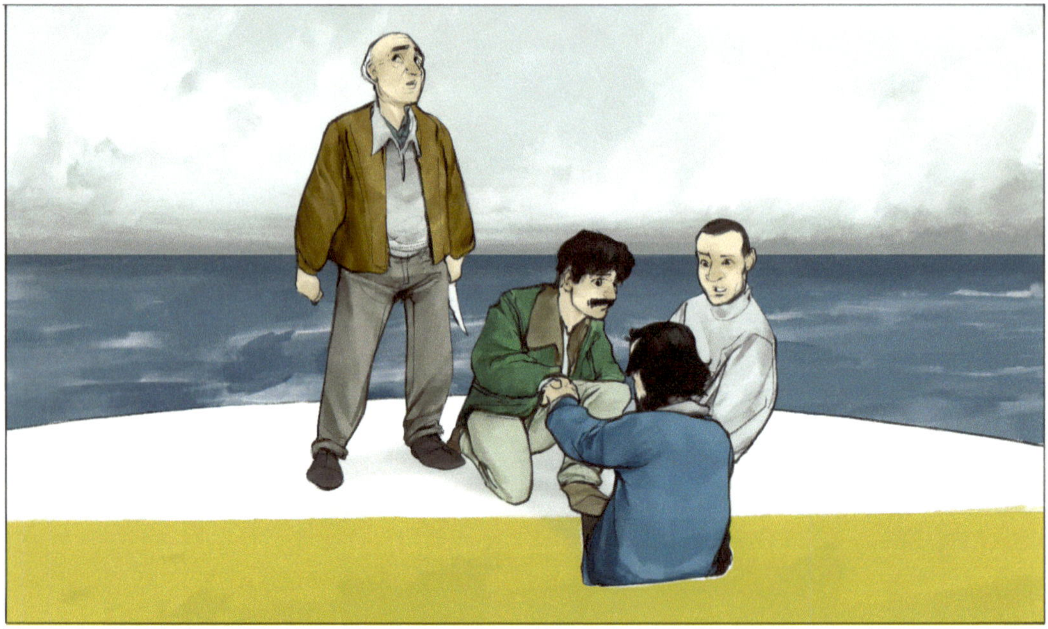

But it wasn't even close to the last.

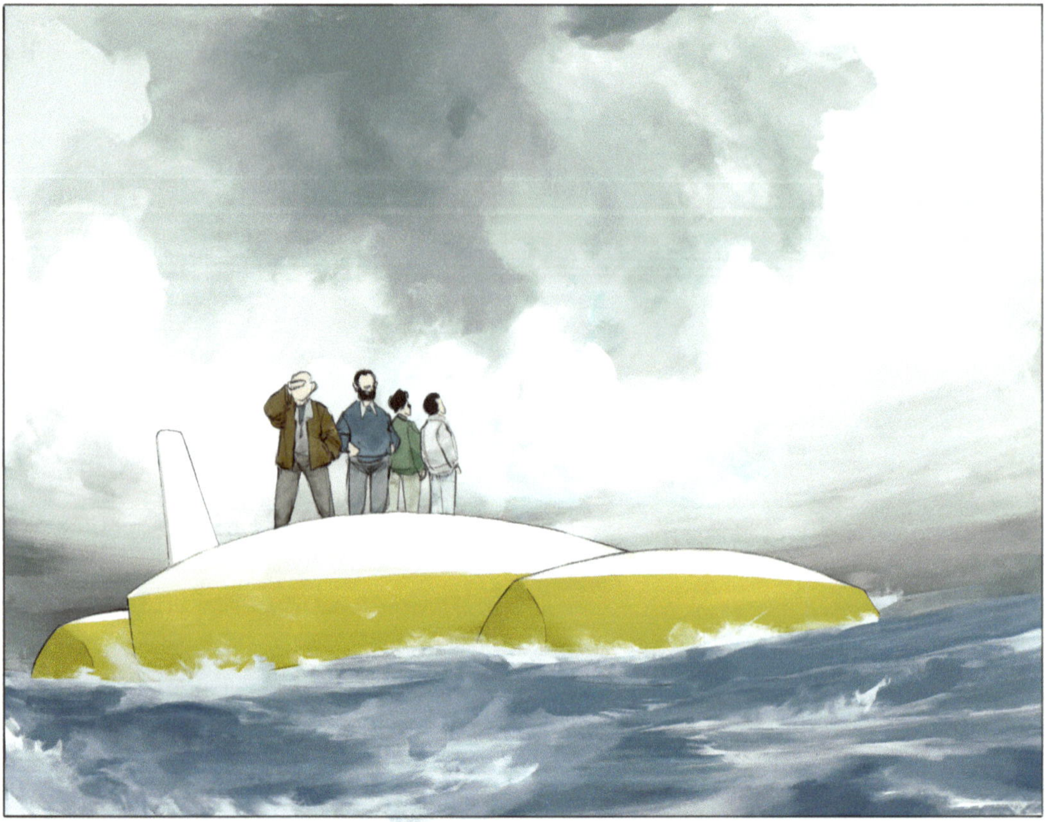

Luckily, the men were prepared.

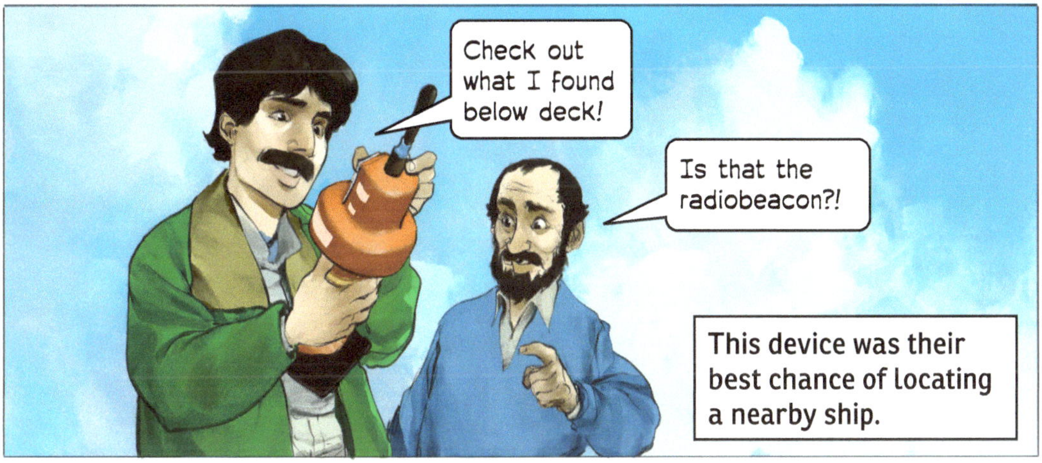
Check out what I found below deck!
Is that the radiobeacon?!
This device was their best chance of locating a nearby ship.

They turned it on and kept it running day and night.

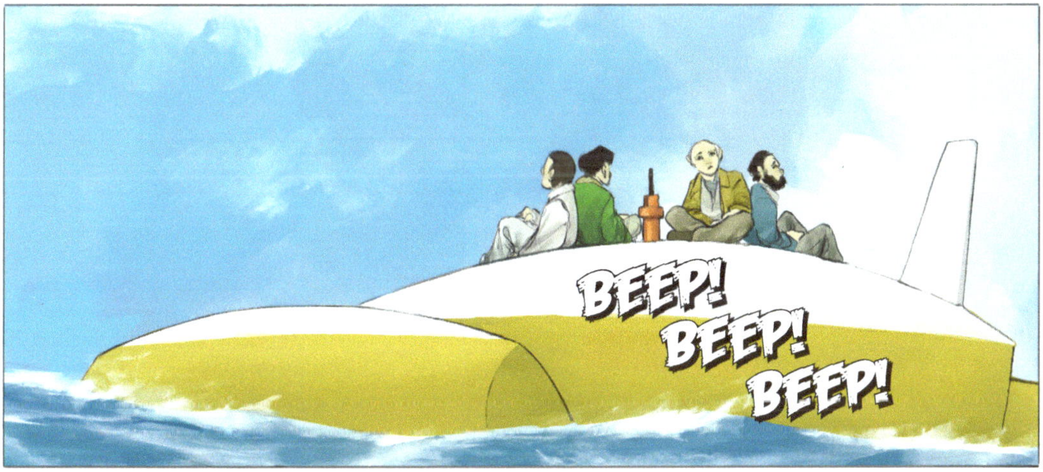
BEEP! BEEP! BEEP!

But it died after eight days.

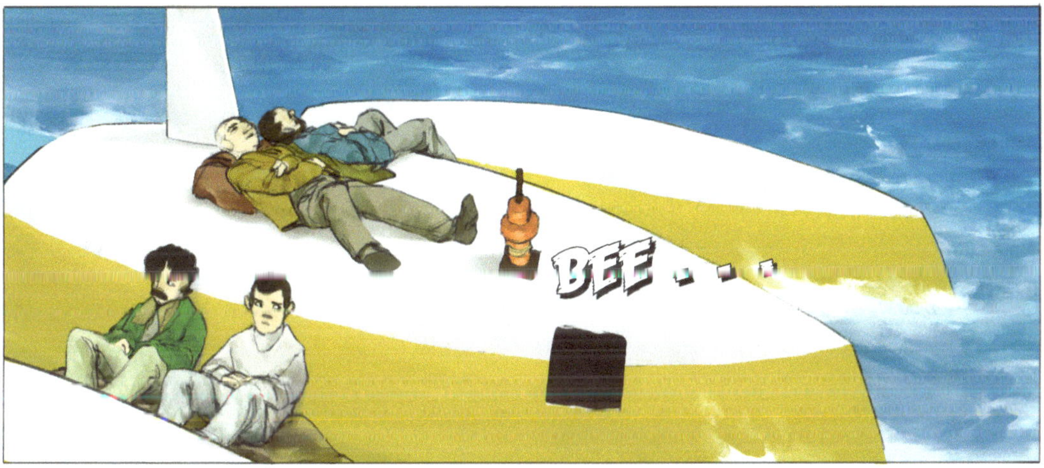
BEE . . .

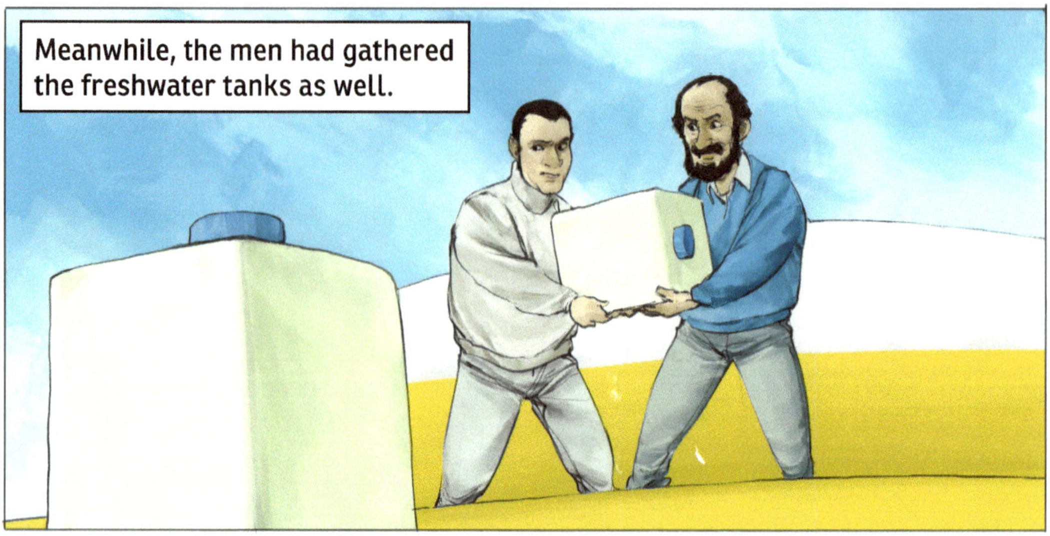
Meanwhile, the men had gathered the freshwater tanks as well.

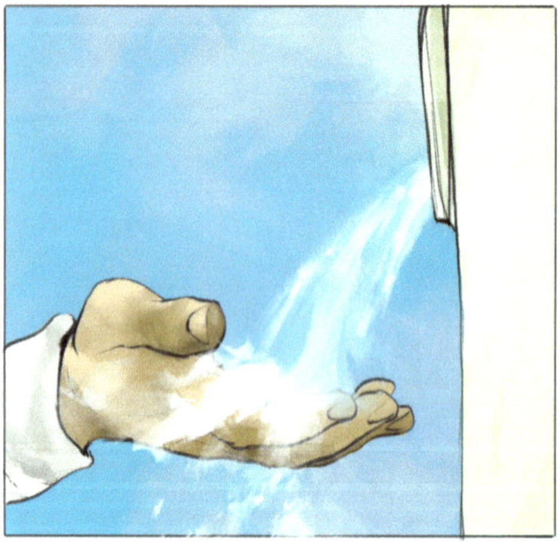

The tanks should have had plenty of water.

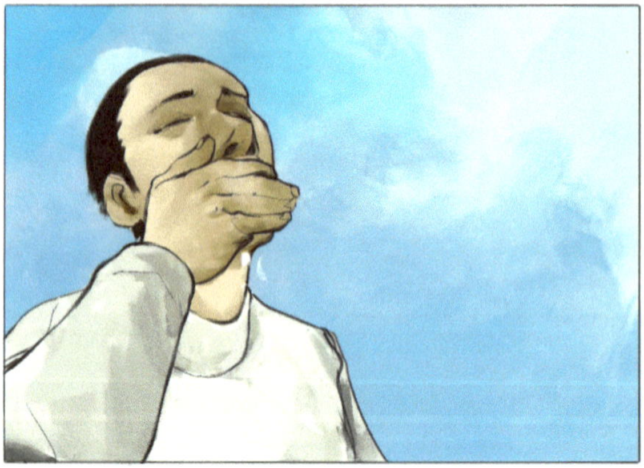

But they had been damaged in the storm.

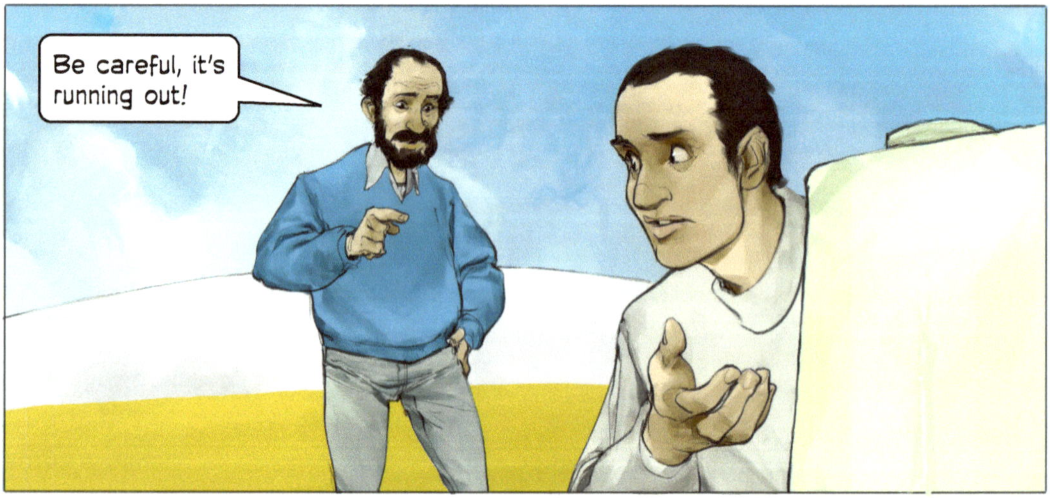
Be careful, it's running out!

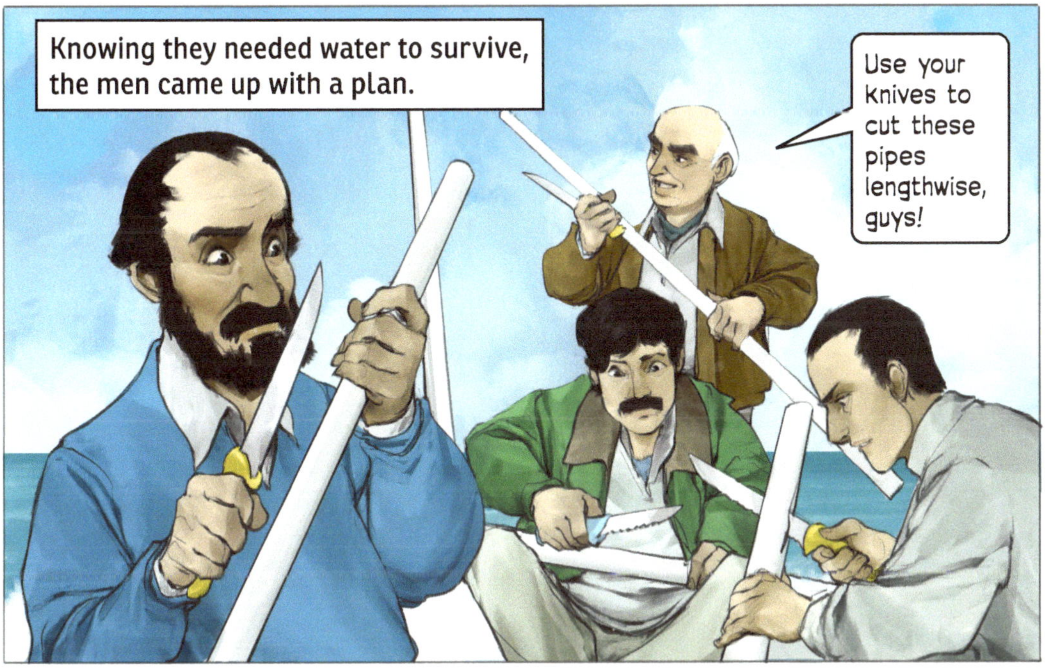
Knowing they needed water to survive, the men came up with a plan.
Use your knives to cut these pipes lengthwise, guys!

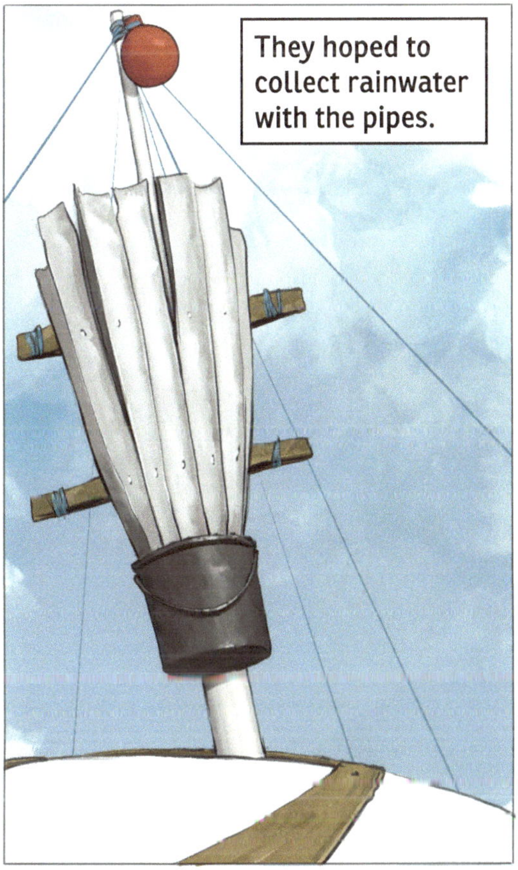
They hoped to collect rainwater with the pipes.

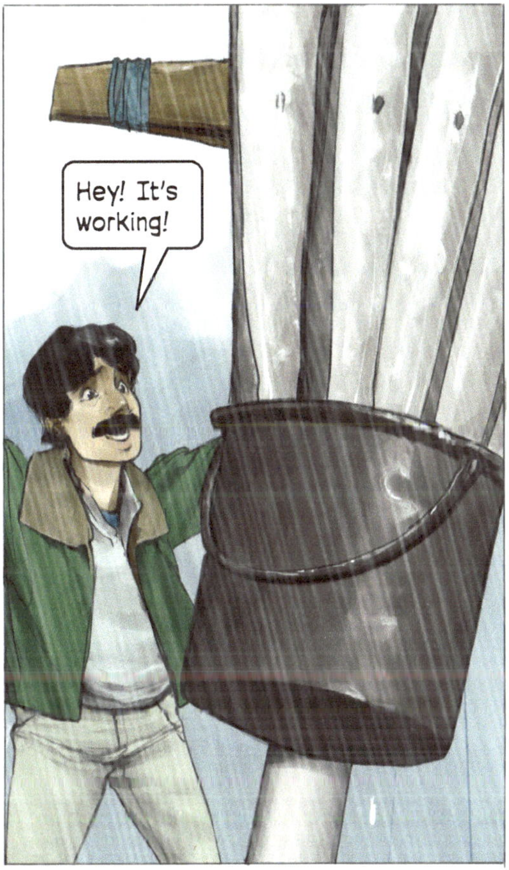
Hey! It's working!

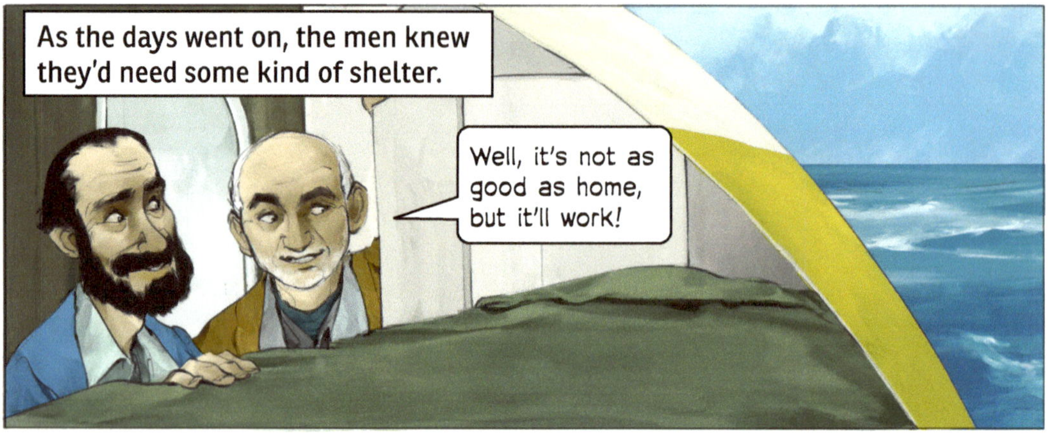
As the days went on, the men knew they'd need some kind of shelter.
Well, it's not as good as home, but it'll work!

The space kept them out of the hot sun.

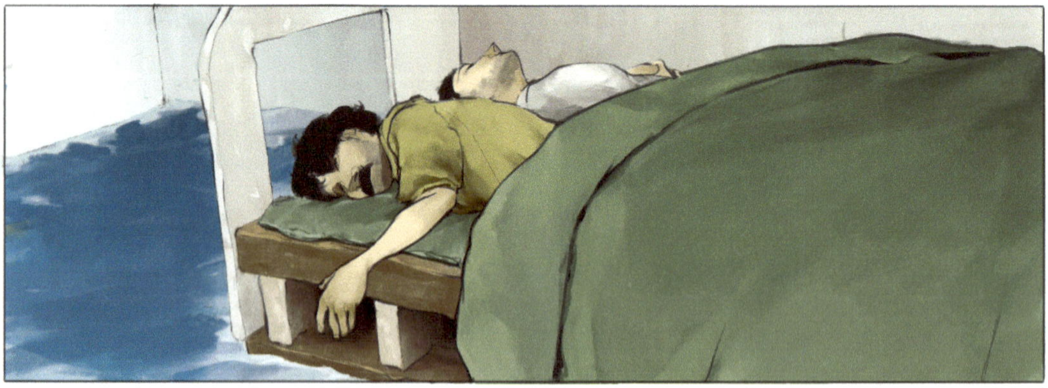

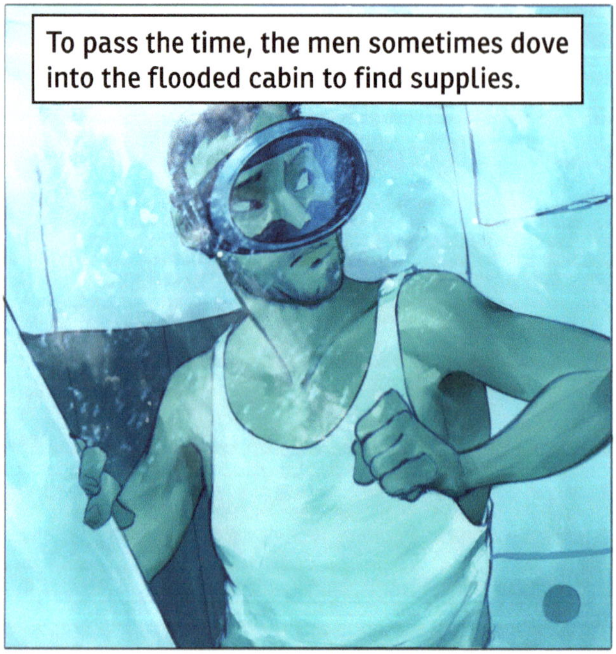
To pass the time, the men sometimes dove into the flooded cabin to find supplies.

They even found kiwi, which gave them some vitamin C.

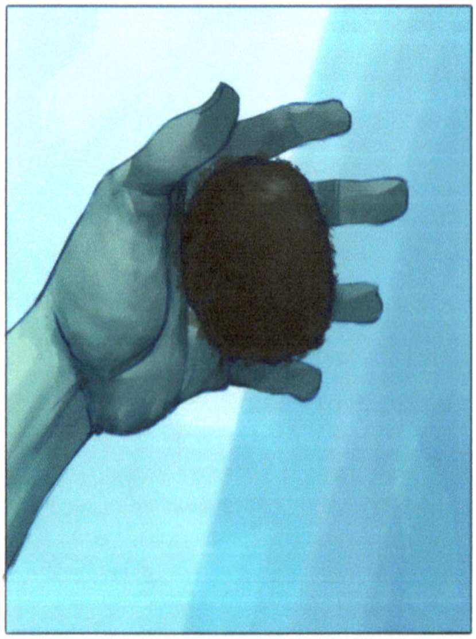

As weeks turned into months, barnacles began to grow along the ship's hull.

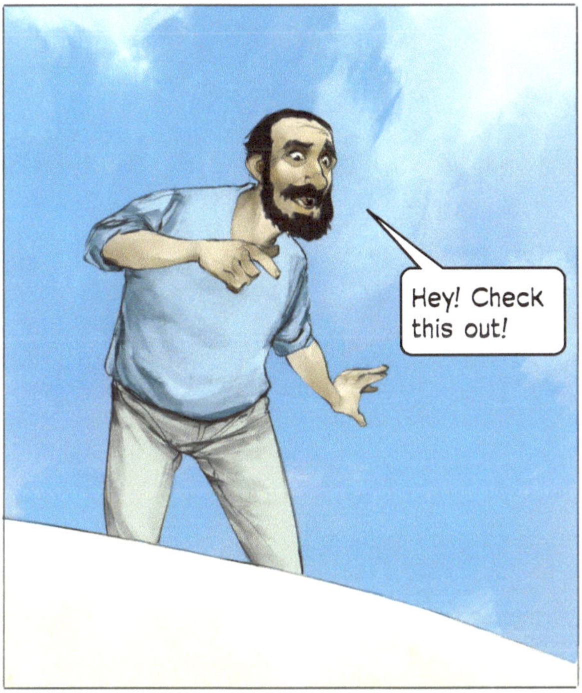
Hey! Check this out!

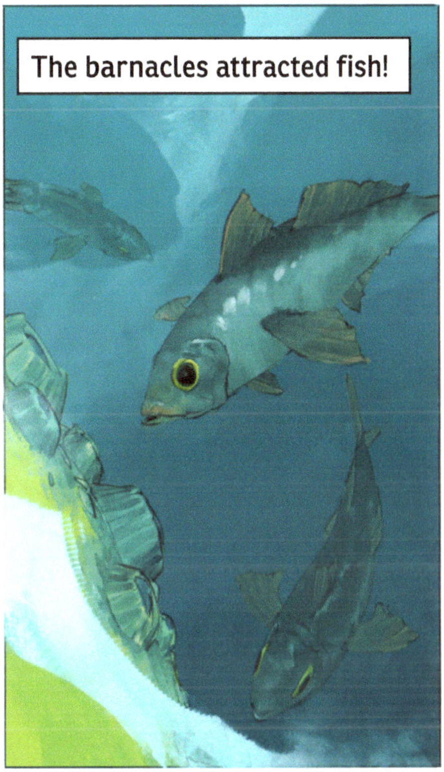
The barnacles attracted fish!

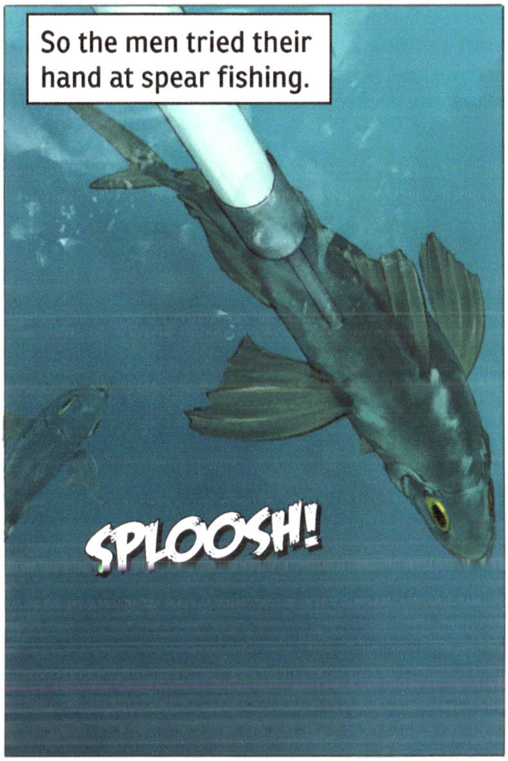
So the men tried their hand at spear fishing.
SPLOOSH!

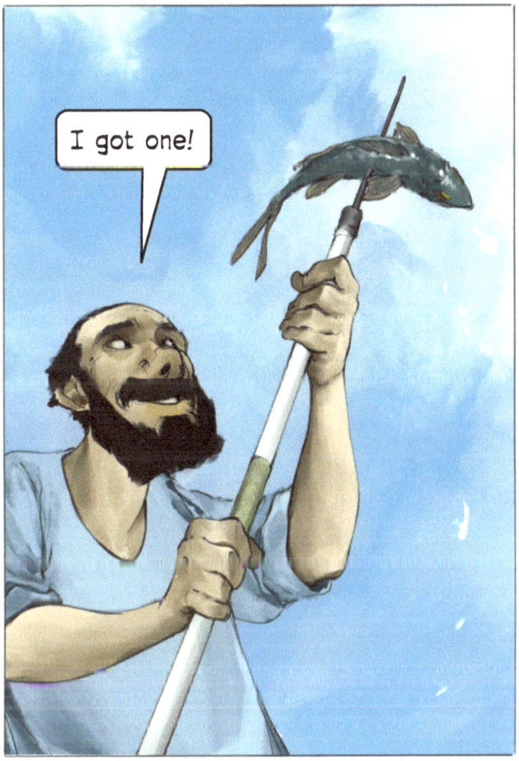
I got one!

Finally, one day . . .

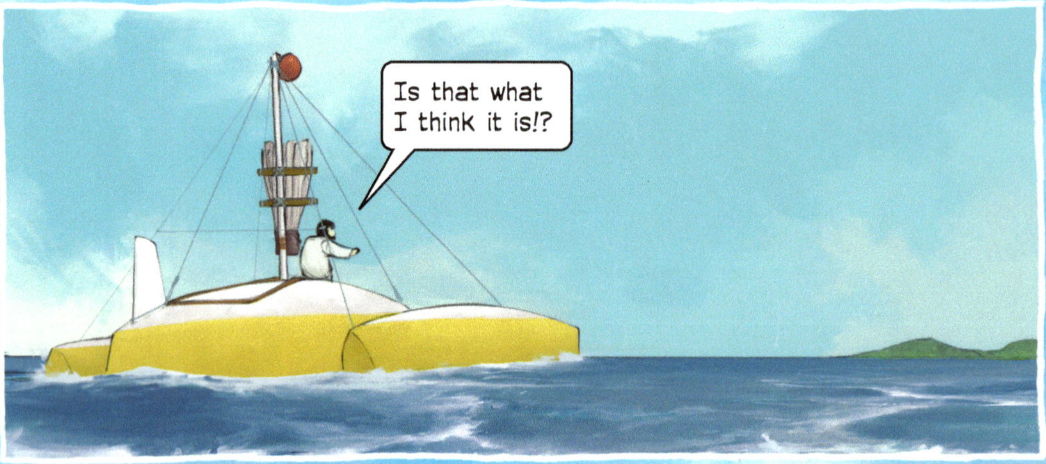
Is that what I think it is!?

The men began to float toward land.

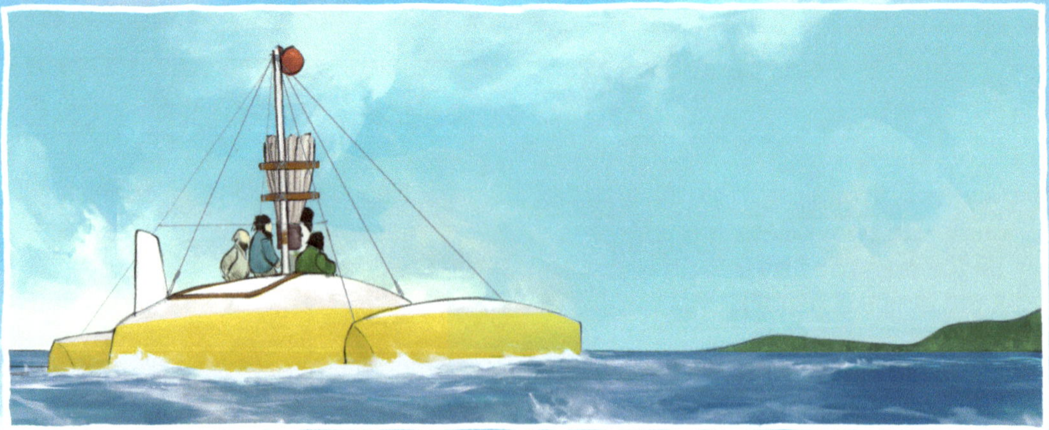

And they eventually hit a reef.

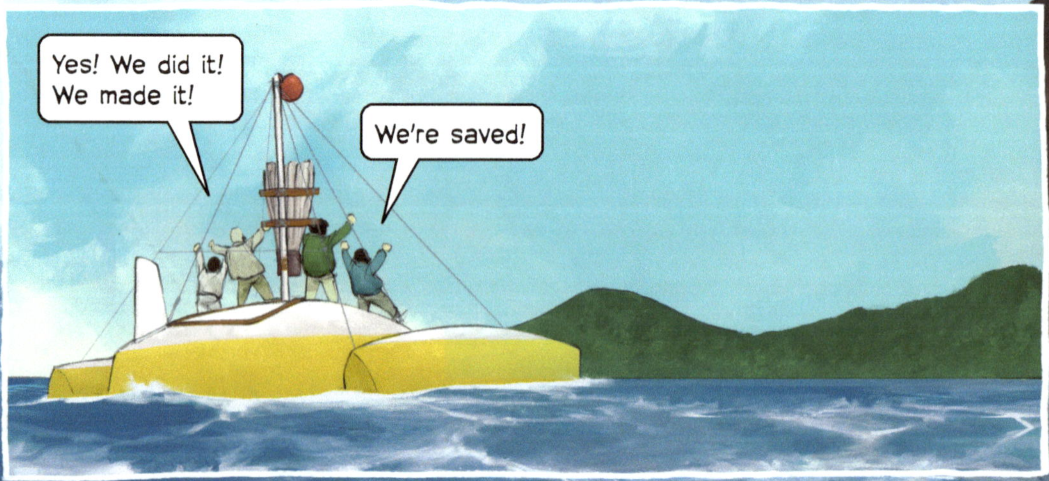
Yes! We did it! We made it!
We're saved!

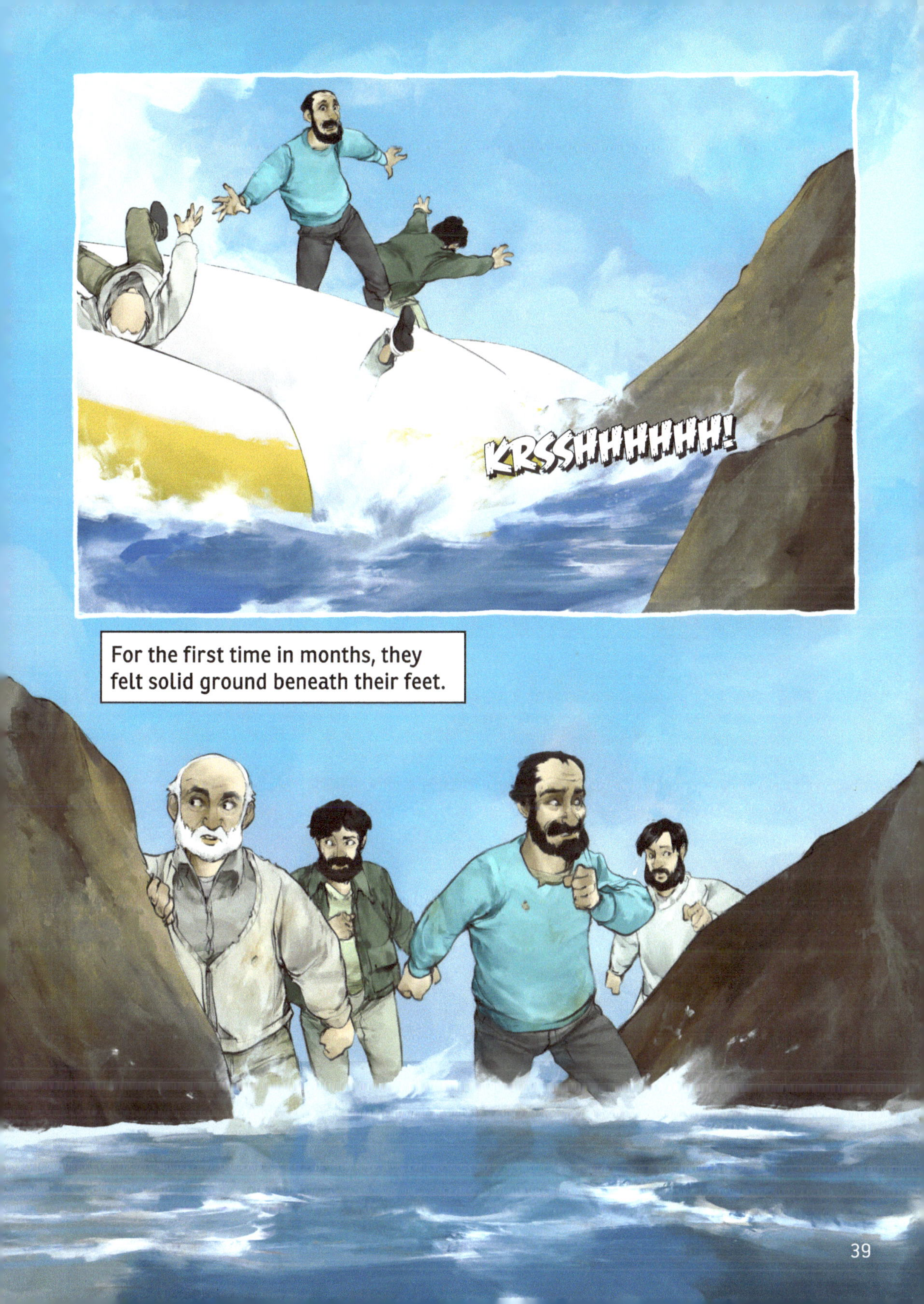
KRSSHHHHHH!
For the first time in months, they felt solid ground beneath their feet.

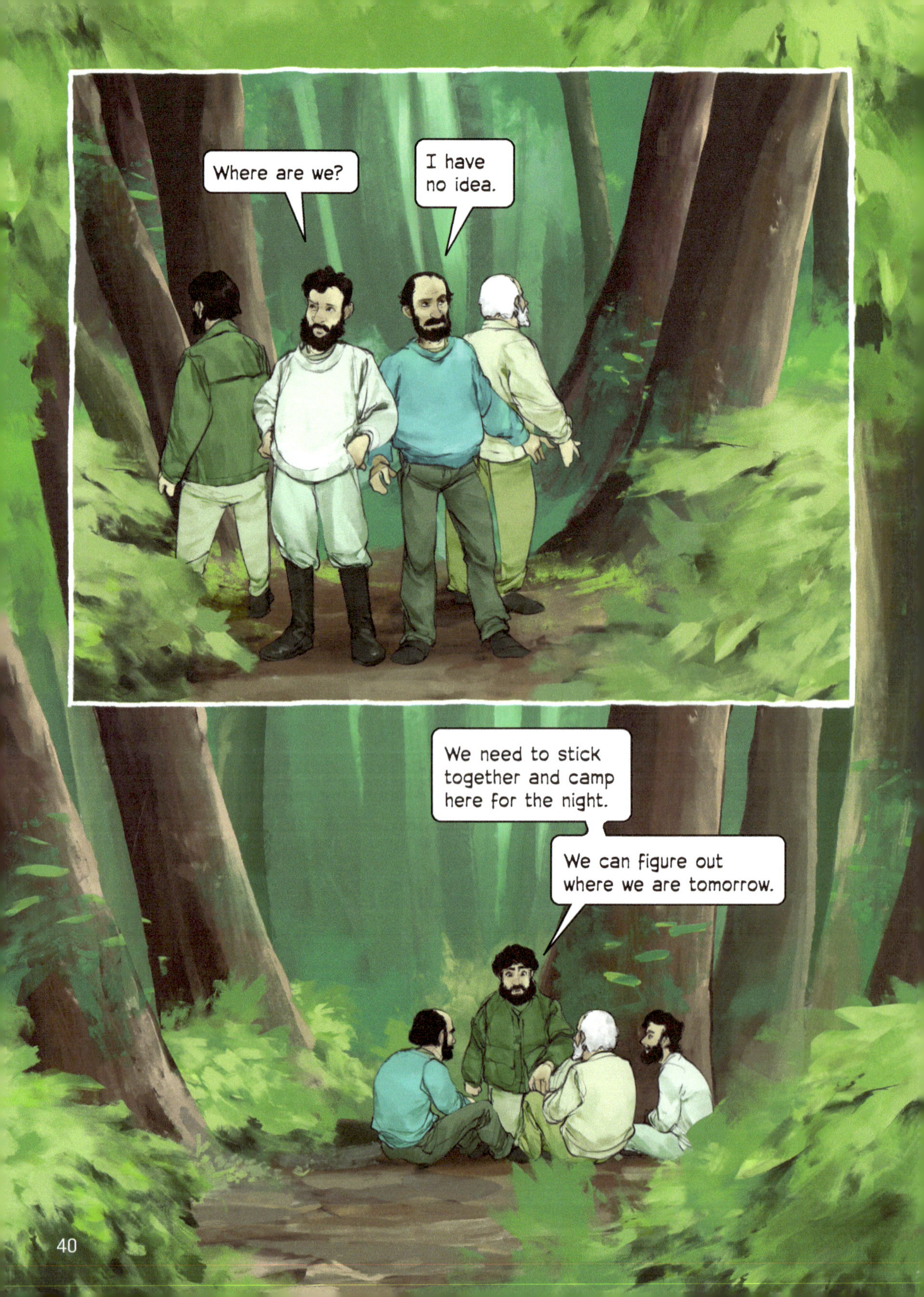
Where are we?
I have no idea.
We need to stick together and camp here for the night.
We can figure out where we are tomorrow.

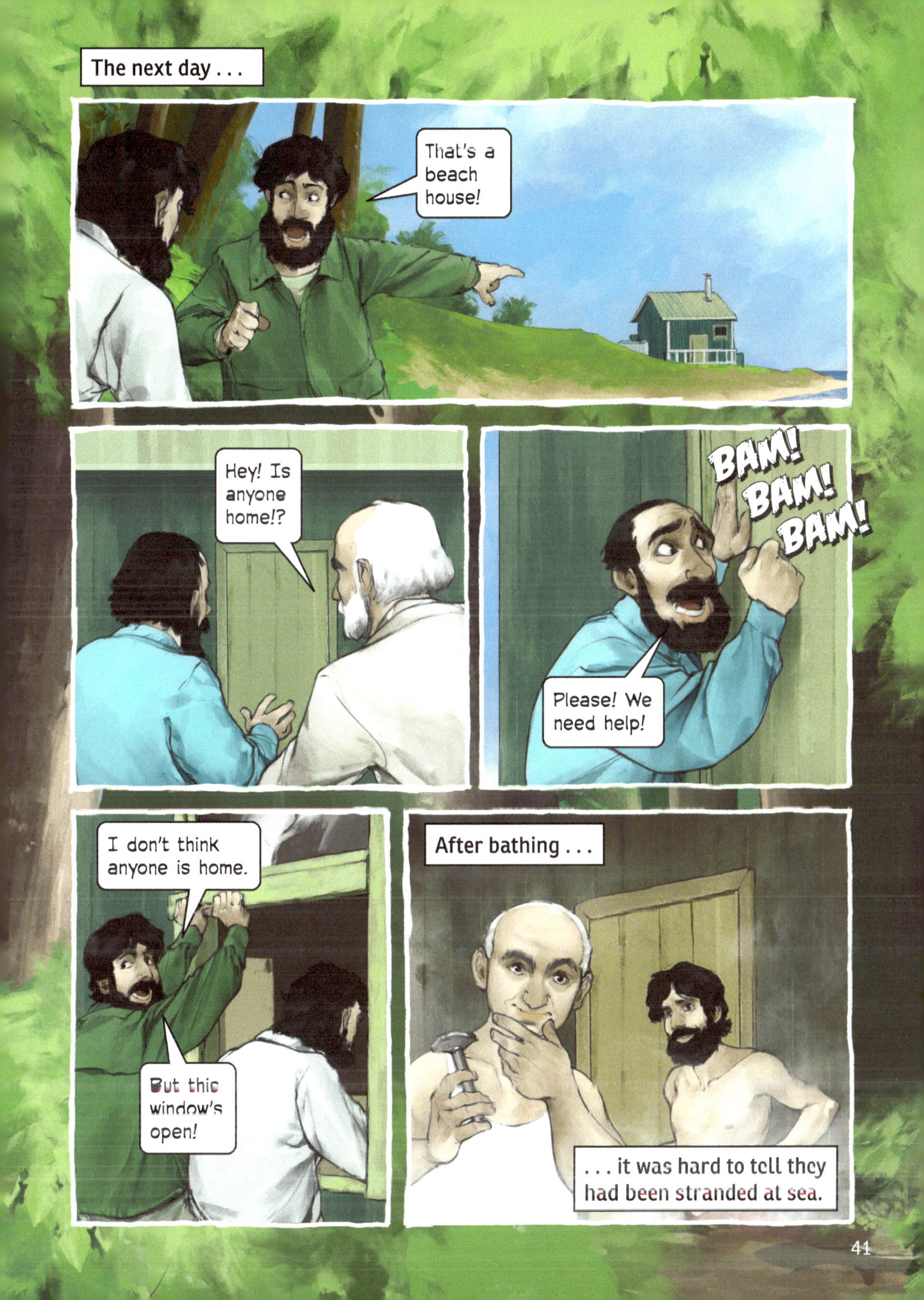
The next day . . .
That's a beach house!
Hey! Is anyone home!?
BAM! BAM! BAM!
Please! We need help!
I don't think anyone is home.
But this window's open!
After bathing . . .
. . . it was hard to tell they had been stranded at sea.

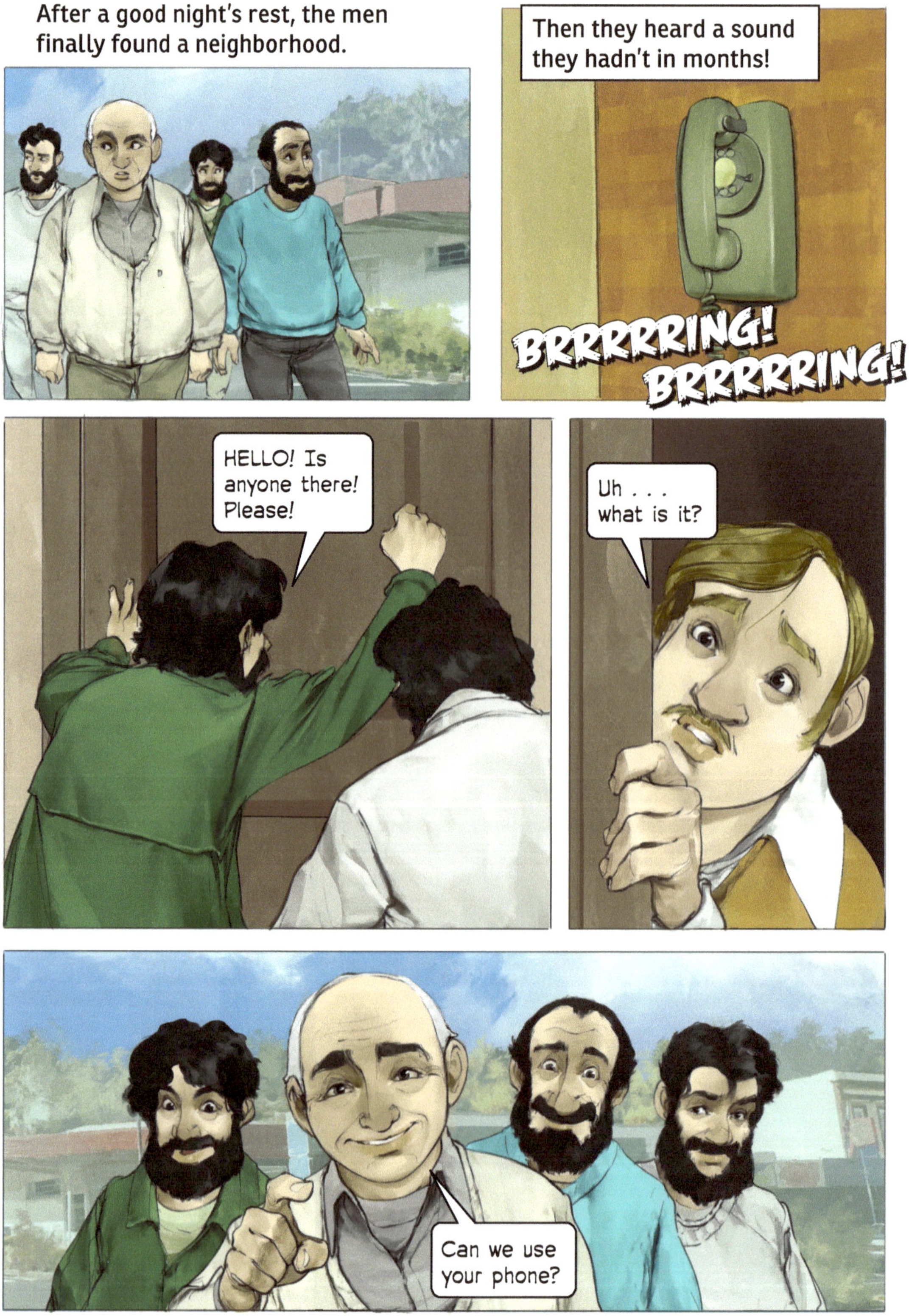
After a good night's rest, the men finally found a neighborhood.
Then they heard a sound they hadn't in months!
BRRRRRING! BRRRRRING!
HELLO! Is anyone there! Please!
Uh . . . what is it?
Can we use your phone?

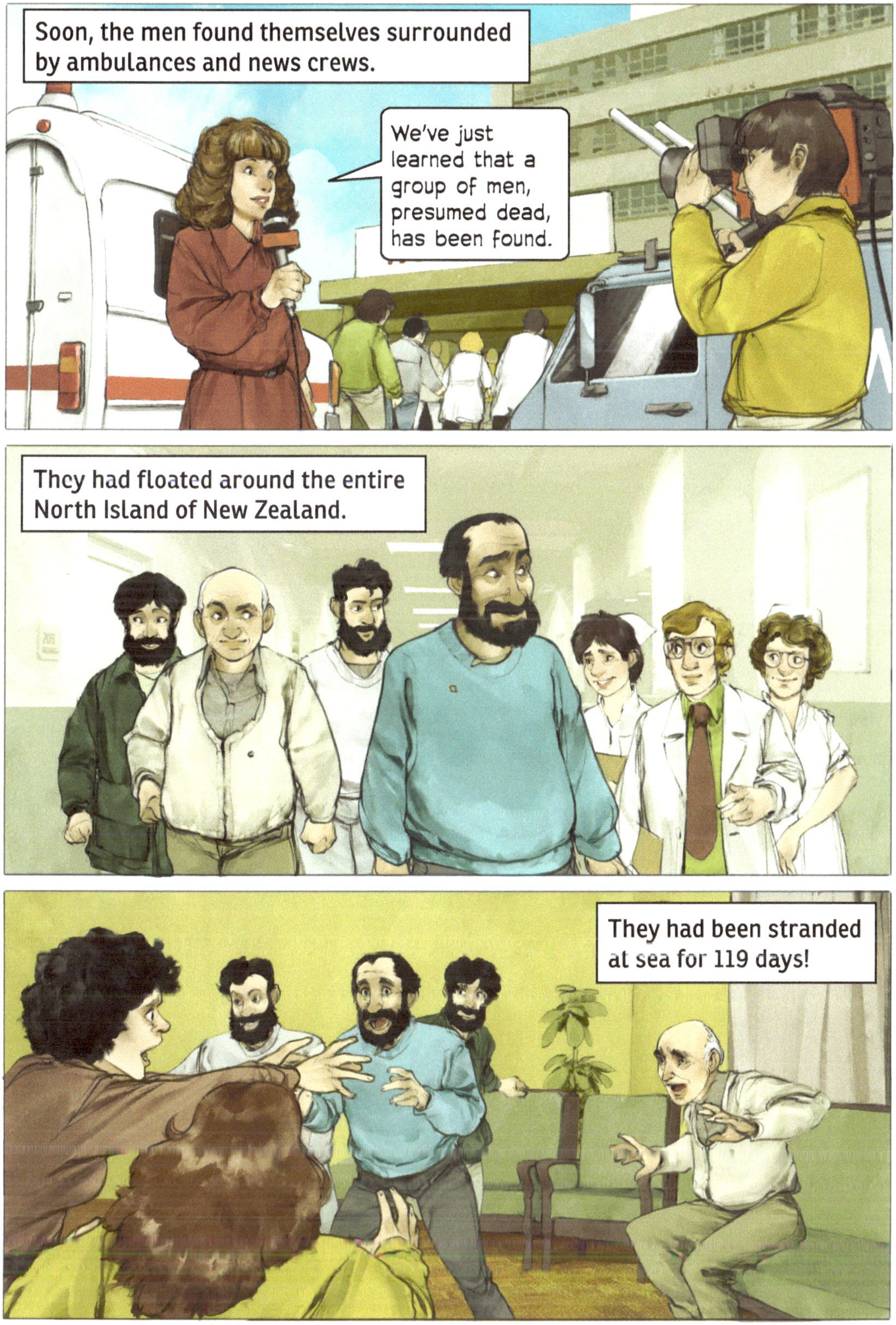
Soon, the men found themselves surrounded by ambulances and news crews.
We've just learned that a group of men, presumed dead, has been found.
They had floated around the entire North Island of New Zealand.
They had been stranded at sea for 119 days!

MORE ABOUT THESE TALES of SURVIVAL

The families of the three boys from Tokelau believed they had died at sea. They even held memorial services for them before they were found.

A book titled *Sea Survival: A Manual* was one of several items Steven Callahan saved from his sinking boat. It was written by Dougal Robertson, who had survived being stranded at sea for 38 days in 1971.

Steven Callahan wrote out the details of his journey in the book *Adrift: Seventy-Six Days Lost at Sea.*

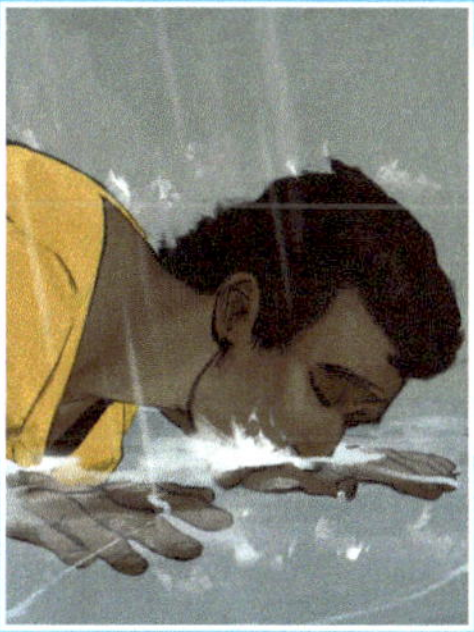

Why couldn't any of the survivors drink sea water? Because ocean water is too salty. While our bodies can handle small amounts of salt, too much salt water will damage our kidneys.

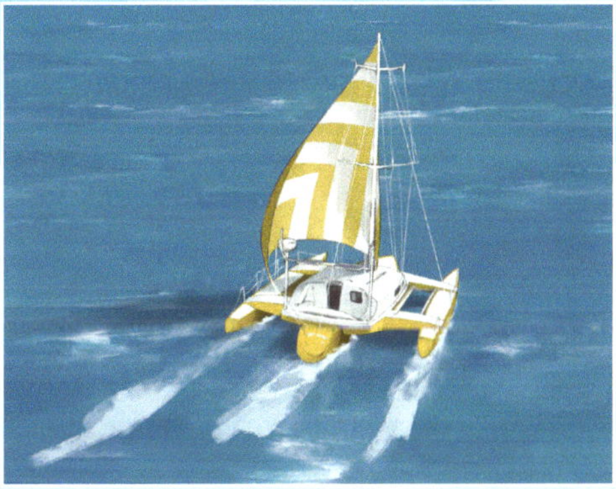

The crew of the *Rose-Noëlle* had their story retold in the 2015 film *Abandoned.*

Some people didn't believe the *Rose-Noëlle* crew's survival story because the men looked clean and healthy when they were found. But no one could prove they had lied.

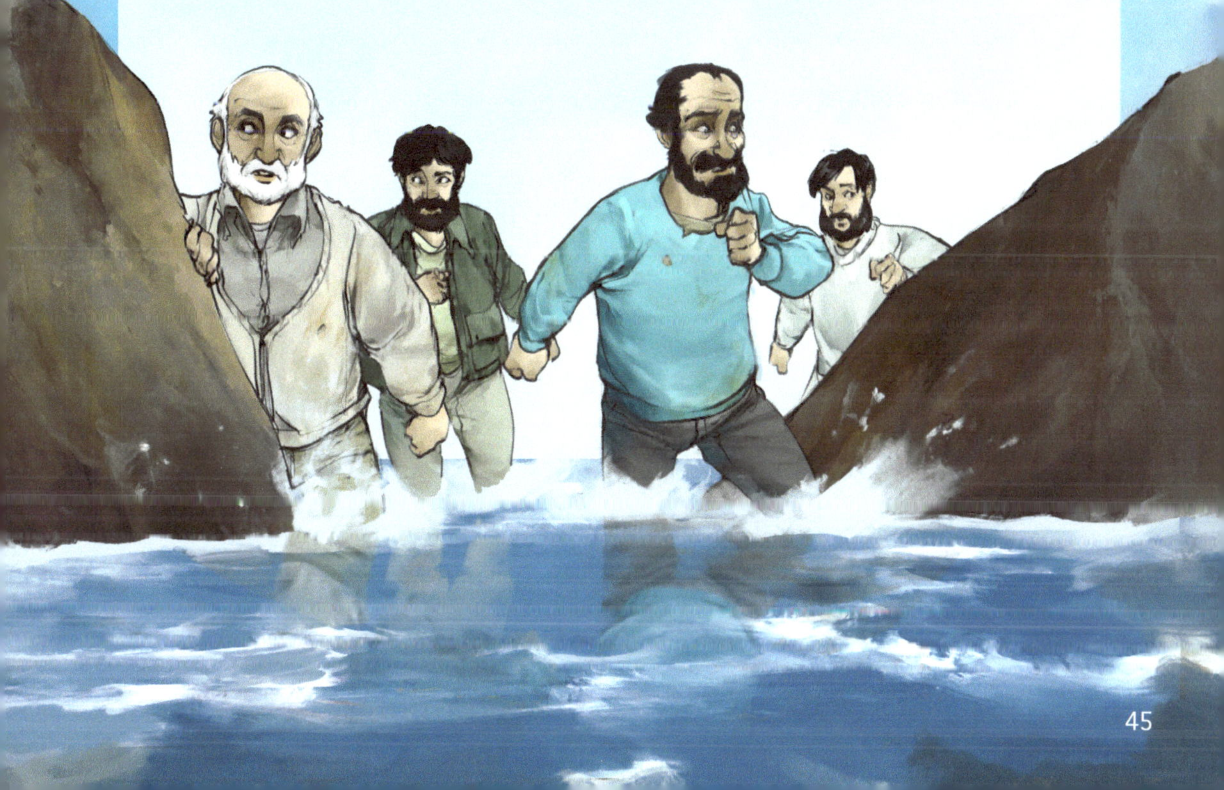

GLOSSARY

barnacle (BAR-ni-kuhl)—a small shellfish that attaches itself to the sides of ships

coconut (KOH-koh-nut)—a fruit with a hard, hairy shell

dire (DIRE)—dreadful or urgent

hatch (HACH)—a hole in a floor, deck, door, wall, or ceiling

kiwi (KEE-wee)—a small, round fruit with brown, fuzzy skin and green flesh

mercy (MUR-see)—the power or whim of something else

radiobeacon (ray-dee-oh BEE-kuhn)—a device that sends out a radio signal indicating a ship's location

rescue (RESS-kyoo)—to save someone who is in danger

routine (roo-TEEN)—a regular way or pattern of doing tasks

vitamin C (VYE-tuh-min CEE)—a substance needed for good health; vitamin C is found mostly in fruits and leafy greens

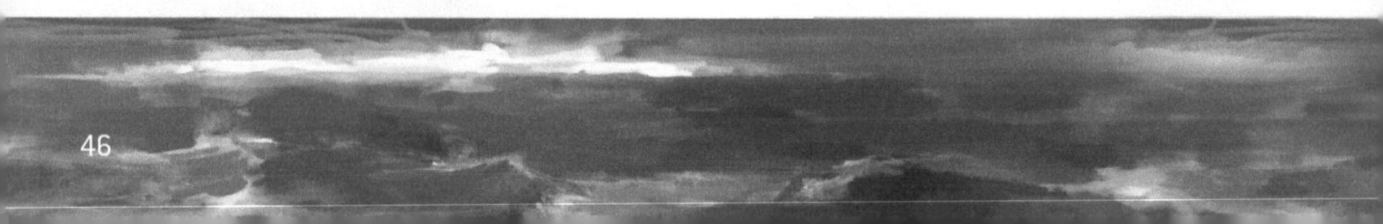

INTERNET SITES

Back From the Dead: The Saga of the Rose-Noëlle
nzonscreen.com/title/saga-of-the-rose-noelle-1996

NBC News: Teens Adrift at Sea Almost Lost Hope
nbcnews.com/id/wbna40390651

Survival at Sea: The Story of Steve Callahan
sailingeurope.com/blog/survival-at-sea-the-thrilling-story-of-steve-callahan

BOOKS IN THE SERIES

TRUE SURVIVAL GRAPHICS
ANIMAL ATTACKS

TRUE SURVIVAL GRAPHICS
DEADLY NATURAL DISASTERS

TRUE SURVIVAL GRAPHICS
LOST in the JUNGLE

TRUE SURVIVAL GRAPHICS
STRANDED at SEA

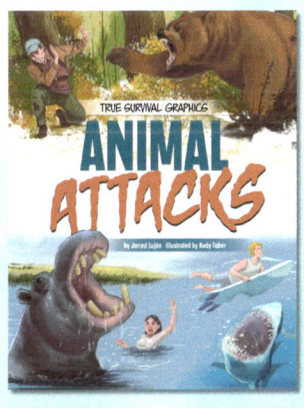

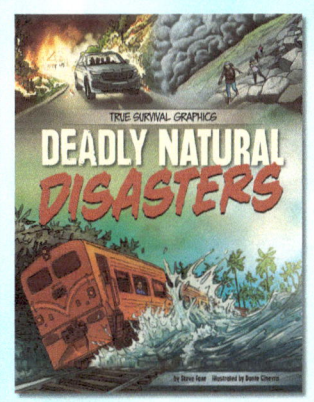

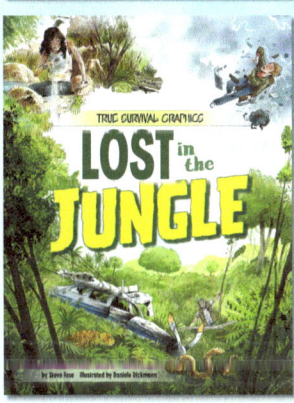

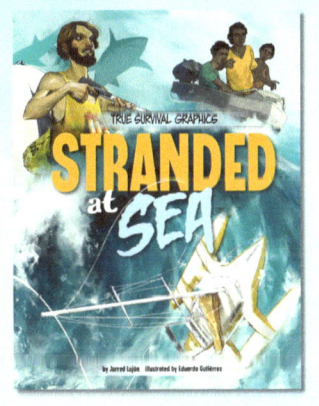

ABOUT THE AUTHOR

Jarred Luján is a Mexican-American comic writer from the borderlands of Texas. He's a 2019 Mad Cave Studios Talent Search winner and a member of the inaugural 2022 DC Milestone Initiative class. Aside from writing, Jarred spends his time trying to convince his cats to be nice to him (his dog remains as loyal as ever).

Photo by Jarred Luján

ABOUT THE ILLUSTRATOR

Eduardo Gutiérrez was born in a little medieval town in northern Spain. He has published some young-adult books in Europe and the USA, and worked for various role-playing games publishers, like Chaosium Inc. He likes vintage stuff, reads about Byzantine Empire and contemporary history, and maintains remarkable diplomatic relations with his three cats.

Photo by @smilekris13